SpringerBriefs in Environmental Science

More information about this series at http://www.springer.com/series/8868

V. Ratna Reddy · Mathew Kurian
Reza Ardakanian

Life-cycle Cost Approach for Management of Environmental Resources

A Primer

V. Ratna Reddy
LNRMI
Hyderabad
India

Mathew Kurian
Reza Ardakanian
UNU-FLORES
United Nations University
Dresden
Germany

ISSN 2191-5547   ISSN 2191-5555   (electronic)
ISBN 978-3-319-06286-0   ISBN 978-3-319-06287-7   (eBook)
DOI 10.1007/978-3-319-06287-7

Library of Congress Control Number: 2014946186

Springer Cham Heidelberg New York Dordrecht London

Printed on acid-free paper

Springer is part of Springer Science+Business Media (www.springer.com)

Contents

Figures

Tables

Chapter 1
Life-cycle Cost Approach: Rationale and Relevance

1.1 Introduction

This volume is about the application of life-cycle cost approach (LCCA) in the management of infrastructure and other investment projects in the context of developing countries. The main objective is to identify potential aspects for its adoption in developing countries with the help of case studies and best practices. It seeks to influence the policy understanding of why life-cycle cost assessment is central to achieving the goals of sustainable development as well as sustainable service delivery and to influence the behaviour of sector stakeholders. The idea is to mainstream LCCA into governance processes at all institutional levels from local to national in order to increase the ability and willingness of decision makers (both users and those involved in service planning, budgeting and delivery) to make informed and relevant choices between different types and levels of products and services.

LCCA can provide 'win–win' strategies in terms of identifying appropriate technologies, products and services that are environmentally, economically and socially sustainable. LCCA prompts policy shifts towards a systems perspective. Adoption of LCCA evolves from life-cycle thinking that needs to be ingrained into macro policy. This calls for awareness building and capacities at the policy and planning levels.

This volume is an attempt towards awareness building among policymakers, researchers and development practitioners about the importance and role of LCCA in achieving sustainable development and provision of sustainable services in the context of developing countries. Specific objectives include:

- To discuss the rationale and relevance of LCCA in the context of developing countries,
- To present the framework and concepts of LCCA,

© The Author(s) 2015
V.R. Reddy et al., *Life-cycle Cost Approach for Management of Environmental Resources*, SpringerBriefs in Environmental Science, DOI 10.1007/978-3-319-06287-7_1

- To discuss real life case studies using LCCA along with best practices, and
- To identify policy challenges for mainstreaming life-cycle thinking at the policy level.

This volume is based on the extensive and intensive meta-analysis of existing literature on LCCA across the world. The focus is on the role of LCCA in attaining sustainable development and sustainable service delivery with reference to developing countries. The volume is organised into three sections. The following section presents the rationale and relevance of LCCA. The analytical framework and concepts are discussed in Sect. 3. Section 4 highlights the policy challenges in mainstreaming LCCA in developing countries and the last section makes some concluding observations.

Developing countries are plagued with poor and fluctuating service delivery with low or no priority for environmental protection. Often these two aspects are interlinked and complement each other in aggravating the problems. The problems are conspicuous in the case of infrastructure-based basic services like water, sanitation, power, health, etc. Main reasons for this include: (1) lack of attention to planning and design; (2) neglect of source protection investments; (3) lack of allocation towards capital or asset management practices; (4) lack of understanding regarding the linkages between different sectors like groundwater aquifers, energy sector; agricultural and household demand for water resources, etc.; and (5) absence of disaster management preparedness or fund allocations towards such eventualities (Kurian and Turral 2010; Reddy and Kurian 2010).[1]

1.2 What Is LCCA and Why?

1.2.1 Background

At the outset, it is necessary to clarify and define the key concepts that are being used. Life-cycle of a product or service is the process from its birth to death. In other words, from extraction of raw material from the natural system to its final disposal (ISO 2006, as quoted in UNEP 2012). All the costs associated with the product life-cycle are considered life-cycle costs (LCC), which has been used traditionally. The term life-cycle assessment (LCA) came into use when environmental impacts associated with inputs and outputs were evaluated. While there is clarity that LCC doesn't include environmental costs, it is not very clear on what components LCA includes. It appears that social life-cycle assessment is not included in LCA. Different terms like environmental life-cycle assessment (E-LCA) and social life-cycle assessment (S-LCA) are used when environmental and social impacts are assessed. Here we adopt the term life-cycle cost assessment or approach (LCCA),

---

[1] For African and Indian experience see WASHCost project publications covering four countries http://www.washcost.info/page/196.

which incorporates all the economic, environmental and social aspects of life-cycle costs; thus, making LCCA the most comprehensive approach.

Life-cycle management (LCM) is the system of management that minimises environmental and socioeconomic burdens of product life-cycle or in the product portfolio of a business organisation. LCM helps make life cycle approaches operational through the continuous improvement of product systems (UNEP/SETAC 2007). Identifying and incorporating the potential environmental impacts into policy is termed life-cycle thinking. This in a way mainstreams environmental aspects into policy. Number terms related to LCCA are in vogue.[2] Important among these are life-cycle inventory (LCI), which is the database and life-cycle approaches that include techniques and tools to inventory and assess impacts.

The experience of developing countries clearly indicates that the focus has been infrastructure provision rather than service delivery. That is, the focus in terms of planning and investments has been confined to production phase to the neglect of pre- and post-production phases. It is observed that expenditure on infrastructure accounts for more than 80 % of the total allocations in rural water supply services (Reddy et al. 2012). This is attributed to the fact that the budgeted unit costs of rural drinking water services do not take source protection or system rehabilitation costs into account. As a result, slippage[3] of service levels has become a regular phenomenon, i.e. service levels deteriorate or fluctuate between full coverage and partial coverage or unsafe resource situations (Reddy and Batchelor 2012). It is argued that unit costs are not only below the required levels but also the composition of costs is biased in favour of infrastructure to the neglect of source protection or natural resource base.

Natural resources, especially water resources, play a critical role in the agriculture-dependent economies of developing countries. The linkages between land, water and energy need to be understood for enhancing the production efficiency of each sector as well as the combined efficiency for enhanced and sustainable food security. In most cases, natural resource systems are being utilised in unsustainable manner in most countries. Their productivities, individually or combined, are very low and vary widely across countries. As a result, these growing economies experience increasing environmental impacts. Fostering sustainable development and mitigating environmental impacts could be possible through following a 'nexus' approach i.e. water, energy and food security. Following the nexus approach would pave the way for achieving 'green economy' (Hoff 2011).

Green economy is the ultimate one that enhances welfare and equity while reducing environmental impacts. This calls for recognising the inter-sectoral linkages and adopting a nexus approach for resource use efficiency and policy coherence rather than following sectoral approaches (Hoff 2011). In the absence of such sectoral integration, resource degradation has been the norm across the sectors,

---

[2] All these terms are included in the glossary.

[3] Slippage is used in the case of water, sanitation and hygiene services (WASH). WASH slippage is defined as the occurrence of a certain level of WASH services that has fallen back in a defined period of time to a lower level of services.

space and time, while socioeconomic inequalities have been perpetuated. Water sector is the most affected in this regard. In the absence of integrated planning and policy coherence between water, energy and food sector, water resources are being over exploited due to distorted energy and food policies. On one hand, subsidies on power, fertilisers and water encourage farmers to use beyond optimum levels (inefficient allocation), on the other distorted output pricing policies often favour high water intensive crops (Reddy 2010). Similarly, subsidised inputs (fertiliser) have promoted intensive agricultural practices resulting in extensive land degradation in India (Reddy 2003).

Promotion of water conservation technologies (WCTs) such as micro irrigation, often takes only the farm level water use efficiency into consideration rather than looking at the watershed or basin scale. It is misleading to conclude that WCTs results in water savings without considering the scale aspects (Batchelor et al. 2014). Net water savings from WCTs at the basin level are much less, than the observed water savings at the farm level as the latter does not take the return flows downstream from flood irrigation. Crop or product profitability needs to take its environmental impacts within and outside their respective sectors. Apart from crop water requirements, methane emissions and contribution to greenhouse gases (GHG) vary across crops. Crop decisions or policies to promote crops need to take these externalities into account (Davis et al. 2008; Gathorne-Hardy 2013b).

In the absence of appropriate water pricing and regulation (economic or social), the extent of recycling and reuse of water has been very limited (Reddy and Kurian 2010). It was observed that water consumption levels vary widely across different bathroom fixers such as flush tanks, faucets, showerheads, etc. (Reddy 1996). Unless one takes the water use (excess) externalities while pricing and taxing these products into account, it would result in unsustainable water use practices. In fact, off late retailers and consumers are also looking for such information for promoting environmentally friendly products (Finnveden et al. 2009).

Perpetuation of distorted and incoherent policies in the context of climate variability has further aggravated the impacts of resource degradation on food security as well as socioeconomic equity. Climate variability has increased the risk and uncertainty in the livelihoods of the farming communities, especially in the rainfed regions. It is increasingly being realised that investment decisions and public policies need to take environmental externalities, negative as well as positive, and the risk analysis into account in order to ensure sustainable development. These observations hold well across the developing world.

Thus, the need of the hour is to formulate policies and make investment decisions addressing environmental externalities that would ensure sustainable services. That is project or programme appraisals need to be more comprehensive in order to move towards green economies. Adopting life-cycle thinking is expected to take care of all these aspects and avoid shifting the burden between sectors and space (UNEP 2012). However, the progress in adoption of LCCA has been limited across the developing world despite the concerted efforts of the United Nations Environment Programme (UNEP) to mainstream LCCA into policymaking over the past decade.

Though European Commission has taken the lead in mainstreaming LCCA into policy, there appears to be still barriers to its broader implementation (EC 2003). Important reasons for this slow progress include: (1) LCCA is data intensive and availability of required data and in appropriate formats is difficult, (2) lack of clarity on drawing a line between what to and what not to include in the case of environmental impacts, and (3) more importantly lack of awareness among policy makers of its adoption and capacities to take up LCCA assessments. Awareness building at the policy level is the main bottleneck, as availability of data is often demand driven i.e. data is generated as per requirements.

1.3 Life-cycle Cost Approach: Rationale and Relevance

Life-cycle Cost Approach (LCCA) is a comprehensive tool that is often used in project evaluation of various investments leading to products or services. Though the basic principles of LCCA are nearly a century old, its systematic use is only about 25–30 years old (Salem 1999). LCCA is an economic assessment or project appraisal tool that can be applied at any phase of the project life-cycle, though it is preferred prior to the investment decisions. LCCA includes the whole chain and spread of activities from the start to end of the product life, termed 'cradle to grave'. LCCA takes a systems approach looking into inter-connectedness and impacts of/and on other related sectors, i.e. including the externalities. Such a systems perspective is valid not only for the environmental dimension but also for social and economic dimensions.

The usage and adoption of LCCA has transformed over the last three decades from a project appraisal tool to an environmental impact assessment tool. During the early phases, LCCA was widely used in infrastructure projects, such as construction, power, etc., for assessing project feasibility studies, affordability studies, source selection studies, repair level studies, etc. (Barringer and Weber 1996; Asiedu and Gu 1998; Korpi and Ala-Risku 2008). During the last decade or so LCCA is being propagated as an appropriate tool for environmental impact assessment and sustainable development (Lundin 2002; Chan 2007; Finnveden et al. 2009; UNEP 2012). Of late, LCCA is being adopted as an asset management tool that can ensure sustainable service delivery (Lundin 2002; Rahman and Vanier 2004; Bloomfield et al. 2006; AAMCoG 2008; Franceys and Pezon 2010; Reddy 2012; Kemps 2012). The evolution of LCCA has also experienced wider adoption across sectors during the last three decades. Initially, LCCA was confined to US defence department for procurement purposes (reducing the operation and support costs), but has now been adopted in various sectors in public as well as private, including construction, transport, manufacturing, energy, real estate, services sector, agriculture, biofuels, etc. (Asiedu and Gu 1998; Jones et al. 2012; LNRMI et al. 2014; Harris and Narayanaswamy 2009; Batchelor et al. 2011; Davis et al. 2008; Gathorne–Hardy 2013a; Iraldo et al. 2014). In fact, United Nations Environment Programme (UNEP) has taken the initiative in 2002 to promote

LCCA by providing a broader and deeper perspective to it. LCCA is being promoted as a tool and method to achieve green economy and to be adopted in various infrastructures and other projects (UNEP 2012).

The wide spectrum of aspects and sectors LCCA is being adopted indicates its potential to deal with number of pertinent policy issues ranging from project appraisal to achieving green economy, sustainable development and sustainable service delivery. Despite its potential to make comprehensive project assessment, its application is often limited to one of three aspects i.e. project appraisal, environmental impact assessment, and asset management (service delivery). And the coverage of life-cycle phases in the assessment is limited (Korpi and Ala-Risku 2008). This is often attributed to lack of data, in terms of quality, to make comprehensive evaluations, especially with regard to environmental impacts (Ayres 1995). Moreover, methodologies for assessing environmental impacts were also limited prior to the 1990s. As a result, studies have been limited to certain phases of life-cycle, such as research and development (R&D), production and construction (production), operations and maintenance (O&M), and retirement and disposal costs (disposal) rather than taking all phases of life-cycle and its interconnected sectors into account. The development of environmental economics during the last three decades has facilitated a more comprehensive use of LCCA. Moreover, LCCA, which has been a production engineer's assessment tool, is gaining acceptance with economists, planners, financial managers and policy makers.

1.3.1 LCCA: Beyond Project Appraisal

Until the beginning of the twenty-first century, LCCA was mainly used as a project appraisal or cost management tool in order to make investment decisions. It is observed that LCC is the most relevant cost management method and LCCA promotes environmental impacts instead of being a pure costing tool (Korpi and Ala-Risku 2008). The increasing concern for environment and sustainable development during the 1990s has provided a new perspective and impetus to LCCA and its adoption. The Rio Summit in 2002 with its clear focus on global green economy has identified life-cycle thinking as a key to achieve sustainable development. That is: 'If the green economy is to bring the necessary changes to guarantee a future for life on Earth, decision making on product sustainability, investment, and policy must be made using life cycle thinking and operationalised through life cycle management, approaches, and tools' (UNEP 2012: 13).

Life-cycle thinking is capable of integrating environmental, social and economic impacts into the decision-making process thus ensuring sustainability in both public and private sector development initiatives. Life-cycle thinking adopts the complete process of a product's life from raw material extraction from the earth to planning, designing, processing, making parts, finished products, their usage and their disposal. In the process it not only takes into account the natural, social and economic resources that are being used in the production but also the impacts, positive and negative, the production process would cause to these resources. Thus, LCCA has the potential to achieve the objectives of nexus and Green economy. Although this is not done often due to complex methodologies involved, the adoption of environmental economic methodologies has facilitated comprehensive LCA i.e. adoption of consequential LCA versus attributional LCA (Finnveden et al. 2009).

Recent studies have shown that different crop systems can be evaluated and compared in terms of water use, energy use and emissions using LCCA. In a study of four different rice production technologies—intensive flooded High Yielding Varieties (HYV), rainfed rice, Systems of Rice Intensification (SRI) and organic rice—were compared for water, energy and greenhouse gas (GHG) emissions (Gathorne-Hardy 2013b). SRI scored high when compared to other rice systems in terms of water, energy and emissions per kilogram of rice produced under the condition of low manure application. While SRI is an environmentally-friendly method with less water and fertiliser requirements, the environmental benefits might get upset if excess manure (organic fertiliser) is applied. Similarly, a comparative assessment of biofuel and fossil fuel production systems using LCCA has estimated that biofuel production has the largest estimated reduction of GHG when compared to fossil fuels (Davis et al. 2008).

1.3.2 Asset Management and Sustainable Services

Another dimension of the LCCA that is less explored is its potential to ensure sustainable service delivery. The use of LCCA throughout the life cycle of an asset or assets appears quite restricted and undeveloped because LCCA is viewed as not necessarily a good budget tool (Barringer and Weber 1996). Lack of full-blown analysis covering all phases of the life of an asset could be one reason, though life-cycle costing in theory includes all costs at various stages of the life-cycle. The adoption of LCCA ought to be broader throughout the economic life of the asset. In fact, LCCA is being used even for economic benchmarking of the assets (Boussabaine and Kirkham 2004, as quoted in AAMCoG 2008).The process helps in monitoring the economic performance of the asset in comparison with expectations set at the beginning of the project.

Such a process helps in maintaining the life of the asset and even extending the lifespan of the systems. This helps in maintaining certain levels of performance, i.e. checking the slippage in services and maintaining sustainability of services. This in the end ensures reduction in system breakdowns, minimises costs, improves system efficiency, financial sustainability and service sustainability. That is getting value for the money invested. It is observed: 'given the restricted budget available for renewal and replacement of assets, there is a need for much greater scrutiny of existing assets in relation to community worth. LCCA can be applied in this decision making process to judge, given the value of an asset to the community, if renewal or replacement is appropriate and when is the optimal time for such an event' (AAMCoG 2008: 13). This also minimises the risk transfers in the case of public-private partnership contracts.[4]

As mentioned earlier, allocations are highly skewed in favour of capital expenditure, i.e. asset or infrastructure creation with least concern for service flows from these investments. While the infrastructure focus is helpful in enhancing the access and productivity in the short run, they have become dead investments with poor and inequitable service delivery in the long run (Reddy 2009; Kurian and Ardakanian 2013). The role of cost components like capital maintenance and resource protection is critical for asset management and sustainable service delivery. These cost components are often given least priority, especially in the public sector provision of goods and services (Reddy et al. 2012). The impact of the imbalance between capital and other recurrent expenditures becomes increasingly critical when coverage rates start climbing. The result is that water supply systems continue to fall out of service as fast as new ones are constructed. Although the approach has gained dominance as a service delivery model in progressively enhancing coverage, recent evidence suggests that there are critical second-generation sustainability concerns.

It is observed in the case of WASH services in four countries (Burkina Faso, Ghana, India and Mozambique) that allocations towards capital (asset) management are totally absent and this is one of the main reasons for the failure of WASH systems (Franceys and Pezon 2010). Even in the absence of allocations, public WASH utilities in India end up spending 5–6 % of the total cost on asset management. As there are no planned allocations, these funds are often drawn from the regular operation and maintenance (O&M) allocations. This in the end affects the up keep of the systems and service levels adversely (Reddy et al. 2012). In the absence of regular capital maintenance or delays in capital maintenance, there will be long periods of service breakdowns or very poor services (Fig. 1.1). And these breakdowns would often result in high rehabilitation and replacement costs pushing the unit costs high. Thus, adoption of LCCA may in fact reduce long run unit costs (allocations) though the initial costs tend to be higher (Reddy et al. 2012).

For instance, in the case of rural water supply systems in India, the observed and normative life of the supply systems were compared across number of agro-climatic zones. The observed life is the actual functional life of the system while the normative life is the technically determined life i.e. the life of the system under normal conditions and maintenance. It is estimated that the observed or actual lifespan of the water supply systems at the aggregate (state) level is 8.2 years versus the normative lifespan of 12.7 years (Table 1.1). While

---

[4] In the case of private-public partnership projects, if the private parties do not include the capital maintenance costs, their total costs would be lower. But when these poorly maintained projects are handed over to the public or the community, the risk of failure becomes high as the adverse impacts of poor or no capital maintenance are realised after a time lag. In this way the risk of service failure is transferred to the public sector or to the communities, while the private party saves on capital maintenance.

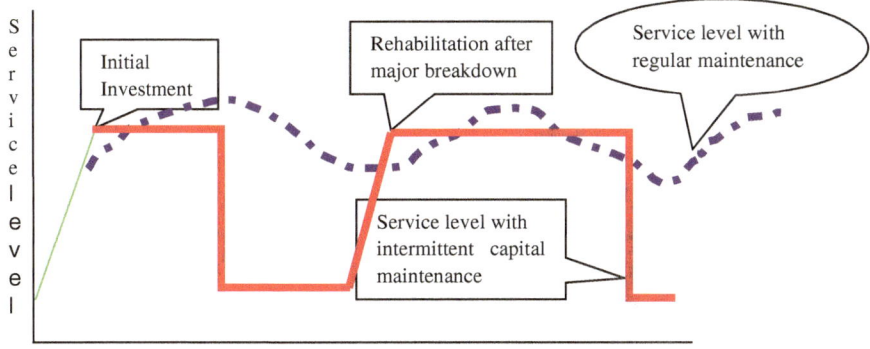

S
e
r
v
i
c
e
l
e
v
e
l

Time / Investment

Fig. 1.1 Capital maintenance and service levels

Table 1.1 Observed and normative lifespans of the rural water systems across agro-climatic zones of Andhra Pradesh

Zone	Observed lifespan[a]			Normative lifespan[b]		
	Average	Range (Min–Max)	CV[c]	Average	Range (Min–Max)	CV[c]
HAZ	7.9	1.0–40.0	69.9	11.2	10.0–30.0	19.1
NCZ	9.8	1.0–49.0	95.6	11.6	10.0–30.0	23.3
GZ	3.7	1.0–31.0	21.9	14.1	10.0–30.0	55.2
KZ	10.9	1.0–49.0	127.9	11.8	10.0–30.0	26.9
SZ	8.4	1.0–45.0	86.9	12.5	10.0–30.0	34.5
SRZ	8.6	1.0–40.0	72.9	13.9	10.0–30.0	56.0
STZ	7.3	1.0–36.0	52.9	13.0	10.0–30.0	44.1
CTZ	7.5	1.0–40.0	54.8	12.7	10.0–30.0	39.4
NTZ	8.4	1.0–40.0	66.8	12.8	10.0–30.0	44.2
AP state	8.2	1.0–49.0	74.5	12.7	10.0–30.0	39.8

Note HAZ High Altitude Zone, *NCZ* North Coastal Zone, *GZ* Godavari Zone, *KZ* Krishna Zone, *SZ* Southern Zone, *SRZ* Scarce Rainfall Zone, *STZ* South Telangana Zone, *CTZ* Central Telangana Zone, *NTZ* North Telangana Zone. More details of these zones in terms of coverage of districts and sample habitations are provided in LNRMI et al. 2014
[a] Estimated using the observed data from the 187 sample habitations spread over nine agro-climatic zones
[b] Based on data provided by the Department of Rural Water Supply and Sanitation, Government of Andhra Pradesh
[c] *CV* Coefficient of variation of the sample habitations in the respective zone
Source Village-wise data collected from the RWSS Department, Andhra Pradesh

the normative lifespan across the zones does not vary much, the observed lifespan varies between 3.7 years in the Godavari Zone and 10.9 years in the Krishna Zone. The observed lifespan could be lower because systems breakdown frequently due to lack of maintenance or due to the hydro-geology of the region (bore well failure). The High Altitude Zone (HAZ) has an observed lifespan

Table 1.2 Functionality of the water supply systems and sources across agro-climatic zones

Zone	Systems (HPs, PSPs, pumps, storage, etc.)			Sources (open and bore wells, tanks, etc.)		
	Total	Functioning	% failure	Total	Functioning	% failure
HAZ	98	95	03	27	21	22
NCZ	164	162	01	36	30	17
GZ	125	74	41	29	8	72
KZ	265	258	03	43	37	14
SZ	189	170	02	70	63	10
SRZ	218	190	13	44	36	18
STZ	358	307	14	92	82	11
CTZ	328	278	15	85	60	29
NTZ	389	339	13	96	62	35
AP state	2,134	1,873	12	522	399	24

Source Village-wise data collected from the RWSS Department at the district level

of 7.9 years, which is close to the state average. The extent of the system and source failure is also the highest at 41 and 72 % respectively in the Godavari Zone, when compared to 12 and 24 % respectively at the state level (Table 1.2). This is mainly due to the quality of water. The turbidity levels in water are quite high in this region, leading to choking of water filters and pumps. Seawater intrusion or salinity ingress is another reason for abandoning the sources in parts of the zone. This clearly indicates that number of factors influence the life of the system and the maintenance of the systems need to be in line with specific natural conditions. Moreover, poor maintenance itself may reduce the lifespan. That is in the absence of allocations towards maintenance and replacement expenditure, service levels may not sustain. Poor maintenance causes major breakdowns resulting in higher unit costs in the end.

1.3.3 LCCA and the Nexus

Nexus is the linkage between water, energy and food security. Supply of water and energy are critical for food security. That is food production is directly influenced by the availability of water (irrigation) and energy is required to access irrigation (groundwater). Source protection and maintenance is critical for sustainable supply of both water and energy resources. LCCA can ensure sustainable services of water and energy through its comprehensive cost allocations that take care of source sustainability, asset management, risk management, pre and post-infrastructure support costs and other externalities (Fig. 1.2). Thus, LCCA ensures source sustainable services through source protection, efficient allocation of resources and risk mitigation at the source level and system level.

Fig. 1.2 Nexus—LCCA—sustainable services

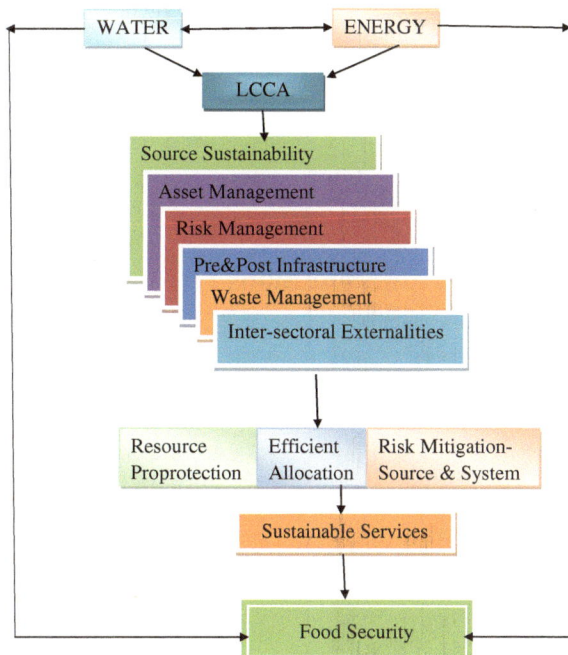

1.3.4 Adoption of LCCA: Scale and Intensity

The adoption of LCCA is widespread covering numerous products in both public and private sectors. Most of the products, however, pertain to manufacturing sector covering construction, energy, transportation, etc. And the purpose of these studies are mostly for design trade-offs (45 %); source selection (38 %) and repair level analysis (13 %) and very few studies have taken all the phases of life-cycle into account while making assessments (Korpi and Ala-Risku 2008). However, this trend has changed since the beginning of this century. As observed above, adoption of LCCA has spread beyond manufacturing covering service sector as well as natural resources. These include water and other natural resources (Koehler 2008; Batchelor et al. 2011; Koroneos et al. 2013); crops (Iraldo et al. 2014; Gathorne–Hardy 2013b) and biofuels (Davis et al. 2008). Of late LCCA is found effective in service sectors like water and sanitation (WASHCost 2010; Jones et al. 2012).

Most of these studies have been framed in narrow life-cycle boundaries thus limiting the potential for achieving sustainable development/green economy goals. There is need for enhancing intensity as well as scale of the LCCA adoption. This calls for policy changes making the adoption of LCCA mandatory at various levels and providing guidelines for achieving green economy objectives. For example, life-cycle thinking is an important element of European environmental policy. A new law in Switzerland requires a complete LCCA of biofuels in order to quantify the fuel tax to be paid (Korpi and Ala-Risku 2008). Adopting life-cycle thinking in all countries, especially in developing countries where environmental protection as well as service delivery is of low priority, is important for achieving cost effective sustainable development. Awareness and capacity building for adopting LCCA methods and tools is a critical step in that direction.

Keywords and Definitions

Cradle-to-grave	A cradle-to-grave assessment considers impacts at each stage of a product's life cycle, from the time natural resources are extracted from the ground and processed through each subsequent stage of manufacturing, transportation, product use, recycling and ultimately, disposal (Athena Institute and National Renewable Energy Laboratory draft 2010).
Environmental aspect	Element of an organisation's activities, products or services that can interact with the environment (ISO 2004).
Life-cycle	'Consecutive and interlinked stages of a product system, from raw material acquisition or generation from natural resources to final disposal' (ISO 2006).
Life-cycle approaches	'Techniques and tools to inventory and assess the impacts along the life cycle of products.'
Life-cycle assessment (LCA)	'Compilation and evaluation of the inputs, outputs and the potential environmental impacts of a product system throughout its life cycle' (ISO 2006).
Life-cycle costing (LCC)	'Life cycle costing, or LCC, is a compilation and assessment of all costs related to a product, over its entire life cycle, from production to use, maintenance and disposal' (UNEP/SETAC 2009).
Life-cycle impact assessment (LCIA)	The phase of Life Cycle Assessment aimed at understanding and evaluating the magnitude and significance of the potential environmental impacts for a product system throughout the life-cycle of the product' (ISO 2006).
Life-cycle inventory (LCI)	'The phase of Life Cycle Assessment where data are collected, the systems are modelled, and the LCI results are obtained' (UNEP/SETAC 2009).

Life-cycle inventory analysis	'The phase of Life Cycle Assessment involving the compilation and quantification of inputs and outputs for a product throughout its life cycle' (ISO 2006).
Life-cycle management (LCM)	'A product management system aimed at minimising the environmental and socioeconomic burdens associated with an organisation's product or product portfolio during its entire life-cycle and value chain. LCM supports the business assimilation of product policies adopted by governments. This is done by making life cycle approaches operational and through the continuous improvement of product systems' (UNEP/SETAC 2007).
Life-cycle management systems	'Management systems that incorporate the basic life cycle principles plus key elements of ISO 9000, ISO 14000 and ISO 26000 to ensure continuous improvement: The plan-do-check-act cycle; Policy, objectives and targets; Procedures and instructions; Monitoring and registration systems; and Documentation and reporting.'
Life- cycle thinking	'Mostly qualitative discussion to identify stages of the life cycle and/or the potential environmental impacts of greatest significance e.g. for use in a design brief or in an introductory discussion of policy measures. The greatest benefit is that it helps focus consideration of the full life cycle of the product or system; data are typically qualitative (statements) or very general and available-by-heart quantitative data' (Christiansen et al. 1997).
Social life-cycle assessment (S-LCA)	'A social and socio-economic life cycle assessment (S-LCA) is a social impact (real and potential impacts) assessment technique that aims to assess the social and socio-economic aspects of products and their positive and negative impacts along their life cycle encompassing extraction and processing of raw materials; manufacturing; distribution; use; reuse; maintenance; recycling; and final disposal' (UNEP/SETAC 2009).

References

AAMCoG (2008) Life cycle cost analysis (LCC Report). The Australian Asset Management Collaborative Group. Retrieved April 10, 2014, from http://www.aamcog.com/wp-content/uploads/2011/08/LifeCycle-Costing-Project-Report-April-2008-(2).pdf

Barringer HP, Weber, DP (1996) Life cycle cost tutorial. Fifth international conference on process plant reliability (Organized by Gulf Publishing Company and Hydrocarbon Processing), October 2–4 (Revised December 2), Marriott Houston Westside Houston, Texas

Batchelor C et al (2014) Do water-saving technologies improve environmental flows? J Hydrol (In Press)

Asiedu Y, Gu P (1998) Product life cycle cost analysis: state of the art review. Int J Prod Res 36(4):883–908

Ayres RU (1995) Life cycle analysis: a critique. Resour Conserv Recycl 14:199–223

Batchelor C, Fonseca C, Smits S (2011) Life-cycle costs of rainwater harvesting systems. Occasional Paper 46. The Hague, The Netherlands: IRC International Water and Sanitation Centre, WASHCost and RAIN. Retrieved April 10, 2014, from http://www.irc.nl/op46

Bloomfield, P, Dent S, McDonald S (2006) Incorporating sustainability into asset management through critical life cycle cost analyses. Water Environment Foundation. Retrieved April 10, 2014, from http://www.environmental-expert.com/Files/5306/articles/13863/497.pdf

Chan AWC (2007) Economic and environmental evaluations of life-cycle cost analysis practices: a case study of Michigan DOT pavement projects. Report No. CSS07-05 (A project submitted in partial fulfilment of requirements for the degree of Master of Science, Center for Sustainable Systems, University of Michigan Ann Arbor, March 22

Davis SC, Anderson T, Kristina J, DeLucia EH (2008) Life-cycle analysis and the ecology of biofuels. Trends Plant Sci 14(3). doi:10.1016/j.tplants.2008.12.006. Retrieved April 10, 2014, from http://www.life.illinois.edu/delucia/Publications/Davis%20Life%20Cycle.pdf

European Commission (2003) Integrated product policy communication. COM 302 final

Finnveden G, Hauschild MZ, Ekvall T, Guinée J, Heijungs R, Hellweg S, Koehler A, Pennington D, Suh S (2009) Recent developments in life cycle assessment. J Environ Manage 91:1–21

Franceys R, Pezon C (2010) Services are forever: the importance of capital maintenance (CapManEx) in ensuring sustainable WASH services. WASHCost briefing note; no. 1b). The Hague, The Netherlands: IRC International Water and Sanitation Centre. Retrieved April 10, 2014, from http://www.washcost.info/page/866

Gathorne-Hardy A (2013a) Baselines and boundaries for rice LCA. International symposium on technology, jobs and a lower carbon future: methods, substance and ideas for the informal economy (The Case of Rice in India), 13–14 June 2013. Organised by University of Oxford & Institute of Human Development, India International Centre New Delhi. Retrieved April 10, 2014, from http://www.southasia.ox.ac.uk/sites/sias/files/documents/Conference%20Book.pdf

Gathorne-Hardy A (2013b) A life cycle assessment of four rice production systems: high yielding varieties, rain-fed rice, system of rice intensification and organic rice. International Symposium on Technology, Jobs and a Lower Carbon Future: Methods, Substance and Ideas for the Informal Economy (The Case of Rice in India), 13–14 June 2013. Organised by University of Oxford & Institute of Human Development, India International Centre New Delhi. Retrieved April 10, 2014, from http://www.southasia.ox.ac.uk/sites/sias/files/documents/Conference%20Book.pdf

Harris S, Narayanaswamy V (2009) A literature review of life cycle assessment in agriculture. RIRDC Publication No 09/029, RIRDC Project No PRJ-002940, Rural Industries Research and Development Corporation, Australian Government. Retrieved April 10, 2014, from https://rirdc.infoservices.com.au/downloads/09-029

Hoff H (2011) Understanding the nexus. Background Paper for the Bonn 2011 Conference: The Water, Energy and Food Security Nexus. Stockholm Environment Institute, Stockholm. Retrieved April 10, 2014, from http://www.construction-index.com/doccbwhollif.html

Iraldo F, Testa F, Bartolozzi I (2014) An application of life cycle assessment (LCA) as a green marketing tool for agricultural products: the case of extra-virgin olive oil in Val di Cornia, Italy. J Environ Plann Manage 57(1):78–103. doi:10.1080/09640568.2012.735991

Jones SA, Anya A, Stacey N, Weir L (2012) Life-cycle approach to improve the sustainability of rural water systems in resource-limited countries. Challenges 3:233–260. doi:10.3390/ch alle3020233

Kemps B (2012) Life cycle costing: an effective asset management tool. Applying LCC contributes to more cost-effective management control of the production facilities of small and medium enterprises (SMEs), Dissertation document (Master of Science in Asset Management Control), International Masters School. Retrieved April 10, 2014, from http://academy.amccentre.nl/thesis/B_Kemps.pdf

Koehler A (2008) Water use in LCA: managing the planet's freshwater resources. Int J Life Cycle Assess 13:451–455. doi:10.1007/s11367-008-0028-6

Koroneos CJ, Achillas CH, Moussiopoulos N, Nanaki EA (2013) Life cycle thinking in the use of natural resources. Open Environ Sci 7:1–6

Korpi E, Ala-Risku T (2008) Life cycle costing: a review of published case studies. Manage Audit J 23:240–261

Kurian M, Ardakanian R (2013) Institutional arrangements and governance structures that advance the nexus approach to management of environmental resources. Paper prepared for Draft White Book: Advancing a nexus approach to the sustainable management of water, soil and waste. International Kick-off Workshop on 11–12 November 2013, UNU-FLORES. Retrieved April 10, 2014, from http://flores.unu.edu/wp-content/uploads/2013/08/FINAL_WEB_whitebook.pdf

Kurian M, Turral H (2010) Information's role in adaptive groundwater management. In: Kurian M, McCarney P (eds) Peri-urban water and sanitation services: policy, planning and method. Springer, Dordrecht

LNRMI et al (2014) Sustainable water and sanitation services: the life-cycle cost approach to planning and management. Earthscan Studies in Water Resource Management, Routledge. ISBN-10: 041582818X|ISBN-13: 978-0415828185

Lundin M (2002) Indicators for measuring the sustainability of urban water systems—A life cycle approach. Environmental Systems Analysis, Chalmers University of Technology, Canada

Rahman S, Vanier DJ (2004) Life cycle cost analysis as a decision support tool for managing municipal infrastructure. NRC Publications Record/Notice d'Archives des publications de CNRC. Retrieved April 10, 2014, from http://nparc.cisti-icist.nrc-cnrc.gc.ca/npsi/ctrl?lang=en

Reddy VR (1996) Urban water crisis: rationale for pricing. Rawat Publications, Jaipur

Reddy VR (2003) Land degradation in India: extents, costs and determinants. Econ Political Weekly XXXVIII(44)

Reddy VR (2009) Water pricing as a demand management option: Potentials, problems and prospects. In: Maria Saleth R (ed) Promoting irrigation demand management in India: potentials, problems and prospects. Colombo, Srilanka: International Water Management Institute

Reddy VR (2010) Water sector performance under scarcity conditions: a case study of Rajasthan, India. Water Policy 12:761–778

Reddy VR (2012) Explaining the inter-village variations: factors influencing costs and service levels in rural Andhra Pradesh. WASHCost-Cess Working Paper No. 22). Hyderabad: WASHCost India and CESS. Retrieved April 10, 2014, from http://www.washcost.info/page/2359

Reddy VR, Batchelor C (2012) Cost of providing sustainable water, sanitation and hygiene (WASH) services: an initial assessment of a life-cycle cost approach (LCCA) in rural Andhra Pradesh, India. Water Policy 14(3):409–429. doi:10.2166/wp.2011.127

Reddy VR, Kurian M (2010) Approaches to economic and environmental valuation of domestic wastewater. In: Kurian M, McCarney P (eds) Peri-urban water and sanitation services: policy, planning and method. Springer, Dordrecht, pp 213–242

Reddy VR, Jayakumar N, Venkataswamy M, Snehalatha M, Batchelor C (2012) Life-cycle costs approach (LCCA) for sustainable water service delivery: a study in rural, Andhra Pradesh, India. J Water Sanitation Hygiene Dev 02:279–290

Salem OM (1999) Infrastructure construction and rehabilitation: risk-based life cycle cost analysis. A thesis submitted to the Faculty of Graduate Studies and Research in partial fulfilment of the requirements for the degree of Doctor of Philosophy, Construction Engineering and Management, Department of Civil and Environmental Engineering, University of Alberta, Edmonton, Alberta, Spring

UNEP (2012) Greening the economy through life cycle thinking. Retrieved April 10, 2014, from http://www.unep.fr/shared/publications/pdf/DTIx1536xPA-GreeningEconomythrough-LifeCycleThinking.pdf

WASHCost India (2010) Cost of providing sustainable WASH services: Experiences from the test bed study areas. Research Report. Centre for Economic and Social Studies, Hyderabad

Chapter 2
Life-cycle Cost Approach (LCCA): Framework and Concepts

2.1 Framework

As discussed, life-cycle cost approach (LCCA) has evolved from a project appraisal tool to a more comprehensive method of incorporating sustainable development aspects in various sectors. LCCA could be conceived in the broader sustainable development framework. The framework consists of three interconnected sustainability dimensions, such as economic, environmental and social. Economic sustainability concept draws from the public finance framework using financial and economic assessment of investments. Environmental sustainability is based on externalities framework (again from 'public good' and public finance). Social sustainability draws from public policy framework where service delivery, governance and social equity are critical. Achieving sustainability on these three counts is a challenge. The nexus approach of water, energy and food security (Hoff 2011) comes close to addressing this challenge. The nexus approach provides a broader framework within which granularity exists. Here, granularity is referred to in the linkages within the sector and sub-sectors, for instance, within water sector, the linkages between surface and groundwater resources, between irrigation and drinking water. Similarly, within drinking water, the linkages between water, sanitation, wastewater, reuse of wastewater, etc., are very much interlinked organically. The granularity is well captured in the three overarching questions raised by Kurian and Ardakanian (2013), (i) intersectionality (critical mass of factors at the intersection of material fluxes); (ii) interactionality (interactions with exogenous factors, viz., policy, economy, environment, etc.; and (iii) hybridity (building transdisciplinary approaches).

Life-cycle thinking is the conceptual idea behind LCCA that reflects the comprehensiveness of the approach in a systems perspective. LCCA takes the whole chain and spread of activities that take into consideration the nexus and

© The Author(s) 2015
V.R. Reddy et al., *Life-cycle Cost Approach for Management of Environmental Resources*, SpringerBriefs in Environmental Science, DOI 10.1007/978-3-319-06287-7_2

the embedded granularity. It takes all the phases of the life cycle of a product or service that are required during pre-production, production and post-production into consideration. These include even the externalities of the production process (Fig. 2.1). It is also argued that the applicability of LCCA in development projects is limited in scope in the context of developing countries, as the all-pervasive social and political drivers are not adequately considered in the present LCCA tools (McConville 2006). LCCA is also data intensive, often making it difficult to use for development work. A life-cycle evaluation of development projects must incorporate diverse factors in a practical manner with a judicious mix of quantitative and qualitative aspects. Further, lack of formal guidelines and reliable past data and difficulty in estimating future costs appear to be the main reasons for the tardy adoption of LCCA. The tool, therefore, must be consistent with successful development practices and simplified for use as a common tool. This could be achieved through a combination of methods and tools for understanding the dynamics.

Though LCCA has potential to deal with various externalities associated with the process, it is not possible to include and assess all the externalities associated with the process of production of any goods and services. While it is easy to scope (consequential) the externalities, it is not easy to assess the impact of these externalities (attributional). It is therefore necessary to define the system boundaries in order to reduce the complexity of assessing the impacts of all the externalities

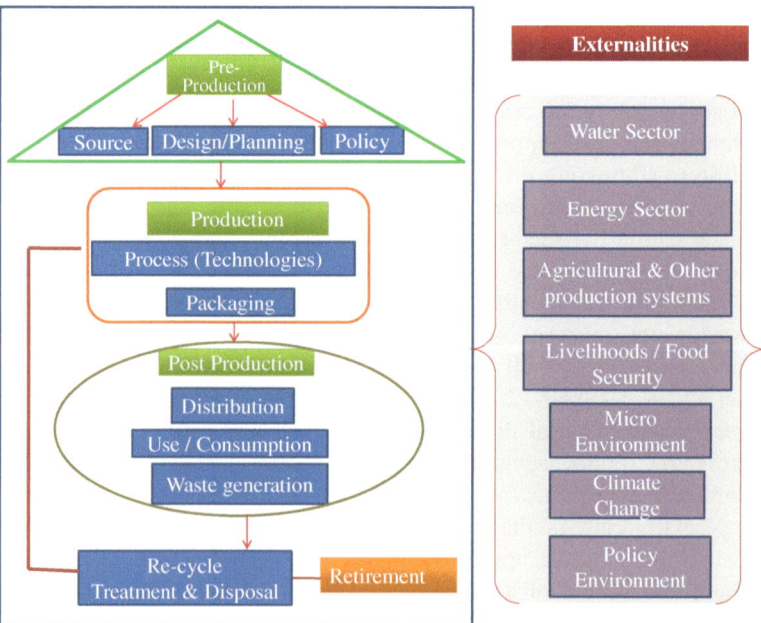

Fig. 2.1 LCCA framework in nexus approach

associated with the process. The choice of system boundaries depends on the nature and type of the product or service in question, which would have important implications on the results (Lundin 2002) and needs to be carefully considered. The life-cycle (or functional) boundaries define the processes to be included in the system, i.e. where upstream and downstream cut-offs are set. Functional boundaries limit the various aspects that are to be included for the assessment. These are mainly related to the environmental externalities. There are three major types of system boundaries: between the technical system and the environment, between significant and insignificant processes, and between the technological system under study and other technological systems (Guinée et al. 2002, as quoted in Finnveden et al. 2009).

Here, we present a generic LCCA framework that shows the possible phases of processes of product or service. These phases could be considered as system boundaries in a simplified version. At each phase, system boundaries can be a set of complex interlinkages. In this generic framework, we look at four phases and the system boundaries (Fig. 2.1). Pre-production phase (level 1) boundaries are defined to ensure resource sustainability and make judicious design and planning for sustainability. The assessment at this level helps in understanding potential environmental issues associated with basic source (raw material extraction). The designing and planning for the production phase is also included and needs to incorporate these costs in conjunction with the policies.

The second phase pertains to production where the emphasis is on infrastructure, technologies and is usually linked to the management agency/institution/organisation. This provides a more complete view of the system in terms of technologies, design efficiencies, planning (viz. linking products and by-products) and packaging. Often the agencies, though aware, are constrained by financial and legislative obligations and tend to override options that allow for a move towards environmental sustainability in the production phase. They either may adopt partially or may not adopt at all. Such a perspective may limit the potential of the agency to identify major environmental impacts or improvements through the life cycle.

The third phase deals with the post-production issues that are often dealt at the community/institutional/household level. These pertain to use/consumption (domestic, agriculture, industry, etc.), and use practices, including waste generation, reuse, recycling, treatment and disposal. This can happen at the production phase as well. And ultimately the retirement of the uneconomic infrastructure. Often, this set gets marginal attention, if not ignored, at the project planning level. This set reflects and determines the adoptability to the system in terms of capacities (technologies), affordability (finance), awareness (quality, health, etc.), attitudes (cultural), etc.

The fourth phase represents the externalities of or to the system that is closely linked and surrounding the main system. The sustainability dimension of LCCA lies in capturing and assessing these externalities. Surrounding systems interact and are critical for the functioning of the core system. Water, energy and land are critical to any production system. While they are often factors of production and included in the costs, these systems also are affected in the production process. Such costs or benefits need to be taken into account. Agriculture production or farming systems (including forestry, livestock, etc.) determine not only demand for the products or services (fertiliser, pesticides, water, etc.) but are also affected in the process (land degradation, chemical use, etc.). These processes would affect the microenvironment in the case of waste or effluent discharge and affect livelihoods positively as well as negatively. Other important factors like climate and policy changes add the risk and uncertainty dimension to the whole process. These need to be taken into account while assessing costs.

This framework can be articulated in the context of water and sanitation that are mostly dependent on scarce groundwater resources in developing countries. Groundwater is exploited for the purpose of supplying drinking water in rural and urban areas. These resources are neither protected from overexploitation nor supported through replenishing mechanisms (like percolation tanks, etc.). There are competing demands for water from agriculture, industry and other livelihoods. In most cases, there are no policies to address these issues. This is part of the pre-production phase, where one has to include the costs of not only identifying and locating the resource but also include costs of planning and design for their sustainable use in the end. During the production phase, different technologies are used to exploit, treat and distribute the water. Here, identifying appropriate technologies that provide optimum benefits is necessary for financial sustainability of the system. Besides, managing the infrastructure is critical for maintaining the life of the infrastructure and sustaining the services. Energy sector plays a critical role at this phase. During the post-production phase, distribution and use are critical for social sustainability in terms of attaining equity in the distribution of service. Here, the institutional and governance aspects play an important role in ensuring social sustainability. Reuse, recycling, treatment and disposal are important for environmental sustainability. Wastewater generated from water, sanitation and hygiene (WASH) services in the urban areas is used for irrigating crops in the peri-urban areas. While the use of wastewater provides livelihoods and economic benefits to communities, it also results in negative impacts like water quality deterioration, health impacts, human as well as livestock, etc. (Reddy and Kurian 2010). Apart from these externalities, the linkages between groundwater and energy also result in externalities such as resource degradation. These externalities can be internalised with judicious planning. The problems of degradation further aggravate in the context of climate variability or policy distortions. Policies like free power would increase the risk of degradation.

In the context of life-cycle costing (LCC), the system boundaries are limited to economic or financial costs. The costs of infrastructure and distribution are only included (Fig. 2.2). Even the use-level costs are not included in the case of financial costs, though economic costs include user costs.

Fig. 2.2 LCC system
boundaries for drinking water
supplies

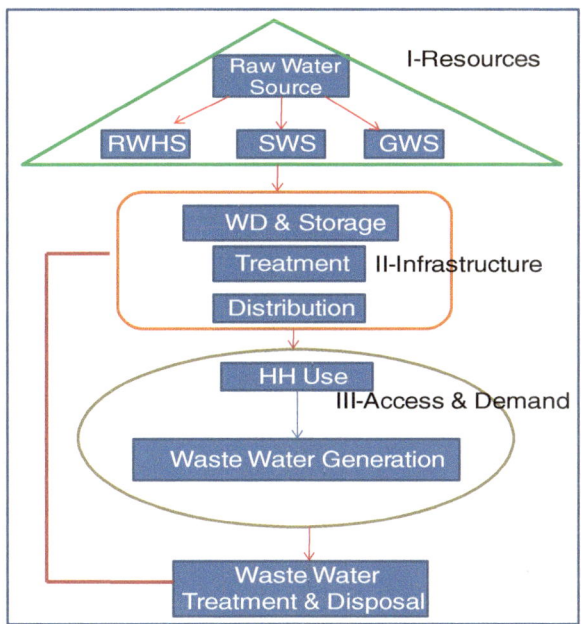

2.2 Cost Components

LCCA analyses the aggregate costs through the life cycle of the system or infrastructure. In a standard LCCA, acquisition costs and sustaining costs are included
at the aggregate level (Barringer 2003). These costs are also termed as recurring
and non-recurring costs or fixed and variable costs. Each of these costs will have
various components of costs at the disaggregate level. Acquisition costs include
hardware and software costs. Hardware costs include mainly infrastructure,
buildings, etc., while software costs include research and design costs, capacity building, etc. Broadly, the cost components include not only the construction and operational costs but also the rehabilitation and information, education
and communication (IEC) costs. These are as follows: capital expenditure on
hardware (initial construction cost) (CapExHrd); capital expenditure on software
(CapExSoft); capital maintenance expenditure (rehabilitation cost or CapManEx);
cost of capital (CoC); direct support costs (ExDS); indirect support costs (ExIDS);
and annual operation and maintenance cost (OpEx). These are broadly grouped
under fixed and recurring costs (Box 1).

   While fixed costs include source protection and construction (hardware) along
with designing and planning (software). Variable or recurring costs include capital or asset maintenance; operation and maintenance costs, CoC, direct and indirect support costs, including training, planning and institutional propoor support.
The delivery of sustainable services also requires that financial systems be in place

in order to ensure that infrastructure can be renewed or replaced at the end of its useful life and to extend delivery systems in response to increases in demand (Reddy et al. 2009).

Depending on the nature of the product or service, it is likely that households, apart from public utilities or private agencies, also invest or incur costs. These costs could be fixed or variable depending on the product or service. It is observed that households often spend substantial amounts towards fixed and variable costs in order to improve the WASH service provided by public agencies, viz. infrastructure such as wells, storage, toilets, etc., and operational costs such as minor repairs, cleaning, etc. These costs are incurred in order to overcome reliability and convenience issues related to water services. Along with these expenditures, households also spend time fetching water and money towards buying water. These are incurred to overcome access and quality problems. While monetary expenditure alone is considered in the case of financial analysis, economic analysis includes both public and household expenditure in monetary terms, as well as opportunity costs. On the other hand, in case of sanitation, public and household expenditure are mutually inclusive, as household expenditure is a necessity and mandatory for construction of household toilets. Hence, both public and household expenditures need to be analysed together for sanitation.

Another set of costs that are important in a comprehensive life-cycle cost analysis (green economy approach) are the costs associated with environmental externalities. These include degradation costs of natural resources like soil, water, air, etc.; emissions or effluents that directly affect livelihoods, health, etc.; and long-term impacts like greenhouse gases (GHGs), etc. These impacts could be positive or negative. They could take place within the sector or product that is being assessed or any other sector linked to the core sector.

Box 1: Cost Components

Fixed Costs
CapExHrd: *this includes government expenditure on infrastructure such as water sources, pumps, storage, filters, distributions systems, etc.*
HHCapExHrd: *this includes household expenditure on infrastructure such as water storage, toilets, wells, pumps, etc.*
CapExSft: *this includes government expenditure on planning and designing costs of the schemes.*
Recurring Costs
CapManEx: *this includes capital maintenance such as rehabilitation of sources, systems, etc.*
CoC: *this includes the interest paid on the borrowed capital for investment in the WASH sector.*
ExDS: *this includes staff salaries, post-implementation activities such as IEC, demand management and training of mechanics.*

ExIDS: *this includes policy planning at the macro level, i.e. central and state.*

OpEx: *this includes regular operation and maintenance of the systems such as energy costs, minor repairs, filtering costs, salaries of water man, etc.*

HHOpEx: *this includes household expenditure on operation and maintenance of water systems, sanitation facilities, etc.*

RTCost: *Retirement costs include the termination costs of the infrastructure.*

Costs of Environmental Externalities

These include resource degradation costs within sector and in other sectors that are linked to the core sector.

2.3 Discount Rates, Annualisation and Functional Unit

All the fixed capital investments are made over the years and are hence cumulated over the years. Similarly, benefit flows are cumulated over the years. When LCCA is adopted at the initial stages of the project, the capital or fixed investments are made in the current year and the recurring investments are made in future years over the life of the system. Some of these costs are regular and expected (operation and maintenance), and others could be irregular and unexpected (capital maintenance). Benefit flows take place in future years. In order to make project appraisals comparable between products or services, all these costs and benefits need to be assessed at the current year. In cases where LCCA is taken up at a later stage of the project, historical costs and benefits are used where costs and benefits would accrue in the past as well as in future. These costs and benefits are inflated to the current year level. Various deflators (future benefits) or inflators (past investments and benefits) are suggested in the literature (Barringer and Weber 1996). These range from the National GDP inflator/deflator (inflation based) to fixed consumption (depreciation) and accelerated depreciation or appreciation. In the case of environmental benefits, lower discount rates are often proposed (Table 2.1).

Different systems have different lifespans, including technical, economic and useful. In order to make the projects comparable, the lifespans need to be standardised by annualising the costs. In order to arrive at the unit costs per year, all the capital costs (CapExHrd) are annualised using the normative lifespans of the systems, i.e. the technical lifespan. Arriving at the lifespan of a system becomes complicated where different components of the system have different lifespans. Using component-wise lifespans for hardware such as boreholes, pumps, pump houses, overhead reservoirs, hand pumps, etc., is more realistic. While normative lifespan is determined technically, it may not hold well in reality. Systems may last longer or shorter than their normative life due to various factors such as poor maintenance and natural factors like hydrogeology; precipitation, temperature and humidity; and natural disasters like floods, droughts, etc. The actual lifespan is the actual number of years the component lasts. By comparing these two, one can

Table 2.1 Present value of investments using GDP inflator for four countries (1961–2010)

	1961	1962	1963	1964	1965	1966	1967	1968	1969	1970	1971	1972	1973	1974	1975	1976	1977
India	100	102	104	109	111	111	113	108	108	115	117	121	130	134	148	160	171
Burkina Faso	100	102	104	109	111	111	113	108	108	115	117	121	130	134	148	160	171
Ghana	100	103	105	113	124	145	157	151	170	188	194	204	235	284	354	458	587
Mozambique																	

	1978	1979	1980	1981	1982	1983	1984	1985	1986	1987	1988	1989	1990
India	203	233	251	273	310	338	356	379	382	358	362	374	393
Burkina Faso	203	233	251	273	310	338	356	379	382	358	362	374	393
Ghana	982	1,702	2,347	3,547	6,230	7,968	17,774	24,050	29,016	41,118	57,237	76,355	97,960
Mozambique	587			100	104	122	138	163	217	244	687	1,019	1,501

	1991	1992	1993	1994	1995	1996	1997	1998	1999	2000
India	400	384	385	379	435	464	466	473	510	533
Burkina Faso	400	384	385	379	435	464	466	473	510	533
Ghana	128,490	154,242	171,440	225,884	293,941	420,468	587,973	702,382	822,128	936,989
Mozambique	2,013	3,242	4,377	6,387	9,946	15,036	24,787	27,011	28,468	29,714

	2001	2002	2003	2004	2005	2006	2007	2008	2009	2010
India	524	545	579	580	603	601	601	623	659	680
Burkina Faso	524	545	579	580	603	601	601	623	659	680
Ghana	1,192,132	1,607,207	1,973,949	2,540,560	2,905,134	3,339,850	6,036,783	7,019,369	8,437,429	9,848,758
Mozambique	33,289	38,242	41,438	43,600	46,858	50,972	55,722	59,836	64,834	68,254

Source: Provided by Mr. Peter Burr, WASHCost Project, IRC, The Netherlands.

assess whether the actual cost of provision is more or less than the estimated costs. Moreover, actual lifespan takes into account the risk and uncertainty associated with the system.

Standardisation is also necessary for comparing the environmental benefits or disbenefits. Functional units are specified for each assessment, and they should be comparable across the products or services, for instance, emissions per unit (kg) of product or wastewater generated per unit of water in filtering (litres).

2.4 Components of Life-cycle Cost Model

The basic LCCA functional form should include the components as indicated in Eq. 2.1.

$$LCC_{xt} = f \left\{ \sum_{t=1}^{n} (CapExhw_{xt}; CapExsw_{xt}; CapManEx_{xt}; CoCap_{xt}; DsCost_{xt}; \right.$$
$$\left. IDsCost_{xt}; OpEx_{xt}) + CoEExt_{xt} \right\} \tag{2.1}$$

where

LCC_{xt} Life-cycle costs of specified product/service
$CapExhw_{xt}$ Capital expenditure on hardware (initial construction cost)
$CapExsw_{xt}$ Capital expenditure on software
$CapManEx_{xt}$ Capital management expenditure (rehabilitation cost)
$CoCap_{xt}$ Cost of capital
$DsCost_{xt}$ Direct support costs
$IDsCost_{xt}$ Indirect support costs
$OpEx_{xt}$ Annual operation and maintenance cost
$CoEExt_{xt}$ Cost of environmental externalities

x represents product or service, and t represents year.

These costs are essential to carry out project appraisal that deals with environmental as well as social sustainability (service delivery) in the short to medium run at least. However, some of these costs are difficult to quantify, especially the costs of environmental externalities. All the costs need to be standardised by annualising the costs. Some of these costs like OpEx are incurred annually, while others need to be annualised. For these investments, past or future, we need to arrive at the present value of these investments in order to make the investments comparable across the schemes. Accordingly, Eq. 2.1 can be written as follows:

$$LCC_{xt} = f \left\{ \sum_{t=1}^{n} pvf_{xt} (CapExhw_{xt}; CapExsw_{xt}; CapManEx_{xt}; CoCap_{xt}; DsCost_{xt}; \right.$$
$$\left. IDsCost_{xt}; OpEx_{xt}) CoEExt_{xt} \right\}$$
$$\tag{2.2}$$

where

pvf Present value factor $(1 + r)^t$
r   Rate of interest or inflator
t   Time period

Rate of inflation or the prevailing rate of interest may be appropriate for estimating the present value or worth. Other alternatives include effective interest
rate (rate of interest–inflation), national GDP inflator could also be used. Once the
whole life costs are estimated, unit costs and annualised costs can be worked out.

2.5 Risk-Based Life-cycle Cost Analysis and Simulations

Some of LCCA components are characterised with risk and uncertainty. Systems
fail randomly and may not follow a time schedule. Time required for rehabilita-
tion/repair and costs may vary. As a result, while normative lifespan of different
systems may not vary much, the actual lifespan varies due to risk and uncertainties
associated with natural factors and unexpected climate events. The risk and uncertainty are often high in the case of products and services associated with natural
resources. The risk factor can be modelled using probabilistic phenomena, that
is by estimating the probability of risk in a particular location due to a particular
event. In the event of risk, the earlier Eq. (2.2) could be written as follows:

$$\mathrm{LCC}_{xt} = f \left\{ \sum_{t=1}^{n} \mathrm{pvf}_{xt} \left(\mathrm{CapExhw}_{xt}; \mathrm{CapExsw}_{xt}; \mathrm{CapManEx}_{xt}; \mathrm{CoCap}_{xt}; \right. \right.$$

$$\left. \left. \mathrm{DsCost}_{xt}; \mathrm{IDsCost}_{xt} \right); \mathrm{OpEx}_{xt}; \mathrm{CoCEExt}_{xt} [\mathrm{Psf}_{xt}] \right\}$$
$$(2.3)$$

where
Psf_{xt} Probability of risk

This formulation is more appropriate in the case of WASH services, as the
dependence on groundwater is quite substantial. In this case, the total life-cycle
cost is modelled as a random variable that is the sum of several cost items. Of
these variables, the CapManEx is a random variable. The randomness or the probability of failure could be estimated using the observed values from the real-life
costing in different agro-climatic locations. These observations can be complemented with expert opinions.

Risk and uncertainty analysis is often carried out using scenario building.
Different scenarios are built using assumptions pertaining to the expected risks.
Scenario building gives a band or range of possible options to choose from, and
simulation models are used to arrive at scenarios. Monte Carlo simulation techniques are used to join probability distributions and economic data to solve

problems of uncertainty using spreadsheet techniques (Barringer and Weber 1996). Monte Carlo simulation techniques use random numbers to generate failure data and cost data considering the statistical distributions. Monte Carlo results are similar to real life because the results have variations around a given theme.

Here is how the Weibull database and Monte Carlo simulations work using the coupling data as an example. Given $b = 2.0$, and $h = 75,000$ h, what is a Monte Carlo age to failure? Solving the Weibull equation for time,

$t = h*\{\ln(1/(1\text{-}CDF/P))\}^\wedge(1/b)$

where CDF is the cumulative distribution function or the probability of failure, which always varies between 0 and 1. The CDF/P range is convenient because spreadsheets also have a random number function, which varies between 0 and 1. This means if the CDF/P = (arrived/chosen by a number between 0 and 1) = 0.3756, then the Weibull age to failure is 51,470 h (or 5.9 years) as driven by the random choice of the number 0.3756. Contrast the Weibull results for age-to-failure with results from the exponential distribution, ($b = 1$) age-to-failure, which produces 35,322 h or 4.0 years using the same random number. When the random numbers are used repeatedly, then specific ages to failure are selected as representative of specific ages to failure. Alternatively, the probability of failure can be estimated using the historical data or expert opinion. Different random numbers or probability scenarios can be modified to build more complex failure propagation tables taking into account how good maintenance practices will reduce the number of failures occurring each period (Barringer and Weber 1996).

2.6 Methods and Tools of Environmental Impact Assessment

While all the relevant life-cycle costs are available in primary or secondary sources or derived from market prices, the cost of environmental impacts needs to be estimated. Various methods have been used to estimate the environmental impacts, as the environmental goods and services are not often available in the market. Here, we discuss some of the important methods used in estimating the impacts.

Methods[1] used in valuations of environmental impacts or costs and benefits can be broadly grouped as direct and indirect. Indirect methods[2] use actual choices made by consumers to develop models of choice for market and non-market

---

[1] We restricted to the methods appropriate for this section. We have not dealt with financial or economic appraisal methods such as Cost–Benefit Analysis (CBA), Cost-effectiveness Analysis (CEA), Multi-Criteria Analysis (MCA), Risk—Benefit Analysis (RBA), Decision Analysis (DA), etc.

[2] These are also known as surrogate market valuation approaches, when information about a marketed good is used to infer the value of a related non-market good.

goods. These include most importantly human capital (HC) approach, replacement cost method, travel cost method (TCM), hedonic pricing method and loss of production method. Direct methods ask consumers their maximum willingness to pay towards a possible change (improvement) in environmental amenities. These methods fall under stated preference techniques where individuals do not make any behavioural changes but state how they would be behaving. The direct methods include contingent valuation method (CVM) and contingent ranking or contingent behaviour. Here, we discuss the methods that are used in assessing the environmental impacts.

2.6.1 Indirect Methods

HC method is most widely used in estimating health-related environmental impacts. The HC approach considers people as economic capital and their earnings as return to investment. Environmental economics focuses on the impact on human health due to bad environmental conditions, and the effect this has on the individuals and society's productive potential. HC approach provides an estimate of direct and indirect burden resulting from the prevalence of disease during a given period. Prevalence of disease-based estimates and present value of future costs are calculated. In the case of incidence of disease, the present value of future direct costs (mortality) and indirect costs morbidity ought to be calculated. There are also non-health sector costs, which are often difficult to estimate due to data limitations. Non-health sector costs include psychological costs such as the influence of mortality on family, life cycles, divorce, widowhood, orphan hood, etc.

Direct health costs are the costs incurred due to mortality and morbidity. Indirect health costs are the value of output lost because of loss of productivity in terms of working or keeping house. Here, the method would estimate the economic costs of illness of a productive human being. Two variants of this can be taken into account while measuring economic costs of illness due to environmental factors: first the cost of medical treatment and second the loss of earnings (working days) due to illness. Together, they provide the total economic loss due to ill health. However, it may be noted that these estimates need to be corroborated with medical science or epidemiological data to correlate the illness with pollution. One way is to conduct laboratory tests of various water samples from the sites in order to check the presence of water-related diseases. The linkages between water pollutants like arsenic and other metals and health hazards are well established; discussions with local doctors help in establishing the linkages between water pollution and the prevailing diseases in the locations. The HC approach provides valuable information, provided its limitations (especially information) are addressed. Though it cannot provide an accurate or complete estimate on the value of life, it does indicate economic costs due to morbidity and premature mortality.

In the context of poor water quality, households adopt various mechanisms. Some buy water, some travel farther to fetch good-quality water, and some knowingly (due to lack of affordability) or unknowingly consume poor-quality water. The last category of households might incur other costs such as medical treatment, etc. In a case study in Andhra Pradesh, India, about 5 % of the households still drink sewerage-contaminated water due to compulsions of non-affordability or non-availability of persons to bring water from nearby town (Kurian et al. 2008). Those who consume the water complain about stomach pain, diarrhoea and joint pains. Women complain that the water quality is getting worse over the years and they are now scared to use the water even for domestic uses. Families consuming this water may have to spend about $5–$8 per month towards doctor fees and medicines. However, there are no serious health complaints of severe sickness leading to loss of working days in the study region. The estimated total costs of water contamination come to $88,763 per year for the entire village (Table 2.2).

The incidence of sickness and unable to work due to pollution was estimated to be between 48 and 50 days in another study in Andhra Pradesh, India, where water was polluted due to discharge of industrial effluents (Reddy and Behera 2006). The effective number of working days lost depends on the probability of getting employment. The average per-year-per-household loss of working days was calculated using the market wage rate in the villages. The estimated average loss per household due to loss of working days was about $28. Number of visits to the doctors before pollution and after pollution and its expenditure revealed a substantial increase. Households in the region used to visit doctors 4–5 times in a year and spent $3–4 on health, but after pollution, it has increased substantially in the affected villages, which has an adverse influence on the socio-economic conditions of the people in the affected villages. Expenditure on health depends on two factors: (a) the severity of diseases and (b) the economic condition of the family. Small and marginal farmers (owning to less than 2 ha of land) visit doctors 20 times per annum, and their expenditure on medical services is $30. However, in

Table 2.2 Health costs of water pollution accruing to households (HH) using various methods in Bommakal Village, Andhra Pradesh, India

Indicator	No. of HH	Economic cost per household/year in US$	Total cost in US$/year
No. HH buying water (averting cost)	20	95	2,000
No. of HH fetching water from town (travel cost)	900	900 @ each HH spends an hour per day in fetching water and the wage rate is US$0.30 per hour (total: $270)	86,447
No. HH drinking contaminated water (human capital)	80	6 (medical expenses)	421
Total	1,000	196	88,868

Source Kurian et al. (2008)

case of medium (owning between 2 and 5 ha of land) and large (owning more than 5 ha of land) farmers, the average number of visits to doctors is 25 and 12 and the amount spent on medical expenses is about $40 and $60, respectively, after pollution. These differences can be attributed to two factors mentioned above.

In the case of China, HC approach was used in the case study of Chongqing region, China. Three components of cost of illness were considered, viz. medical treatment, loss of work and premature death. Three diseases were linked to contaminated water (i.e. hepatitis, dysentery and selected cancers) and were taken into account for estimating costs. In the case of premature death, loss of earnings during the working age (18–60) due to death was estimated. Median age of the patients to die is estimated as 53 years; hence, the cancer patient loses seven years of working life. Individual contribution to production is estimated using the per-capita growth (8 %) and a discount rate of 12 % were used. The total loss due to health damage was estimated to be $21.7 million when HC approach was used (Yongguan et al. 2001).

Averting Costs (*AC*) approach states that in order to avoid the damage due to environmental degradation, one has to spend some money. For example, the victims of environmental damage replace their environment by moving away from the affected area. The costs, which the victims incur by moving to a clean/healthy environment, are called averting or replacement costs. One of the techniques adopted in the averting cost method is that of direct observation of actual spending on safeguards against environmental risks. In the context of health impacts, households may either treat the water on their own (filtering, boiling, etc.) or switch over to bottled water in order to avoid adverse health impacts due to drinking of unclean water. Data pertaining to the AC are based on the households' actual spending on treatment of drinking water and purchased bottled water from the market. In a study of Andhra Pradesh, India, it was observed that about 2 % of the households buy water from the market in order to avoid adverse health impacts. The estimated cost of this averting behaviour is estimated at $95 per household per year (Table 2.2). In the arid regions of Rajasthan, almost a quarter of the total households buy water from the market and spend more than $1 per day per household (Reddy 1999). Water treatment costs are estimated at about $0.9 million in the Chongqing region of China.

TCM uses peoples' actual behaviour and hence captures the actual use values. Travel cost models are based on an extension of theory of consumer demand, with specific reference to value of time. This method, which is the most straightforward of the indirect methods, recognises that visitors to a recreation site pay an implicit price—the cost of travelling to it, including the opportunity costs of their time. Though this method is often used to estimate the willingness to pay for the facilities of a site using information on time people spent on getting to a site, a modified version of this method can be used to estimate the value of time. This method (random utility theory approach) is based on the assumption that the households' source choice decisions depend upon at least two sets of explanatory variables: (i) source attributes, which affect the households' utility, and (ii) household characteristics, which reflect difference in tastes and preferences. According to random utility theory, the probability that household 'h' chooses alternative source 'j' equals the probability that the utility derived from using source 'j' is greater than the utility derived from any other alternative. Following this framework, utility function is estimated with two sets of variables, i.e. source attributes and household characteristics.

The functional form is as follows:

Uih f [Xih, Z1h]

where,

Uih Utility derived by household 'h' using a source site 'l'. Here, utility is indirectly determined by the choice of the source [site].

Xih represents source attributes like distance between source and household, time spent, money paid for collecting water, etc.

Zih represents household characteristics like income, social status, education level, preferences, etc.

In this model, the dependent variable (source/site) is a dichotomous variable, and hence, it is estimated with the help of conditional logit model.[3] This model is found to be useful in estimating the household's value of time and hence suitable for adaptation in the context of valuation of resources. Two clear cases of such adaptation are drinking water and fuel wood where rural households spend substantial amounts of time in fetching/hauling them. In this (conditional logit) model, the value of time spent by the household is given by the ratio of the two coefficients measuring time and money spent for water or fuel wood by the household. Here, the value of time is defined as the marginal rate of substitution between the time spent in collecting water/fuel wood and money paid[4] for them. Health costs of using poor-quality water can be estimated if households have access to two sources with different source characteristics in terms of quality, time spent/money spent. For instance, the extra effort put in/amount spent by households for collecting/buying good-quality water is the value households place on health. One problem that may arise here is the existence of markets for these items. It may be difficult to find markets for drinking water and fuel wood in all the regions, especially drinking water. Another problem[5] here may be the large variations in tastes, availability of alternative sources, incomes, etc., which can be taken care of with appropriate econometric techniques. On the whole, TCM is believed to be a useful tool and found to have worked well in different contexts.

---

[3] Conditional logit is used to deal with the data structure, which includes both groups of independent variables—source attributes vary across sources while household characteristics do not vary across sources.

[4] This is calculated in terms of the price times the quantity of water/fuel wood consumed per day. In other words, the values of water/fuel wood if purchased at market price (even if the household is not actually purchasing them in the market).

[5] Various other problems related to Travel Cost Method are not considered here.

Greater proportion of households resort to fetching water from far-off places to avoid the ill effects of poor-quality water, which is mainly due to their poor economic status in the developing countries. Villagers go to nearby towns to fetch water from municipal supplies. Studies observed that households spend an hour to bring two cans of water (about 40–50 l) and in summers, it becomes worse as wait times can last 2–3 h (Kurian et al. 2008). About 80 % of the households resort to this mode, and the estimated costs are $270 per household per year. In arid regions, the travel time tends to be substantial, i.e. 18 h per day per household. Often, these travel costs are not accounted as the opportunity costs of labour, especially for women and children (the main fetchers of water), tend to be zero in some rural areas of developing countries (Reddy 1999).

Hedonic pricing method uses surrogate markets to impute values of non-market goods. It estimates the implicit price of the non-market characteristics, which differentiate closely related or explicitly similar products. This method is widely used to value environmental amenities or disamenities associated with a good, using market values (i.e. property valuation approach or wage differential approach). For instance, take two units of houses, which are identical in all respects except one, i.e. air pollution. Their prices would differ if people place value on clean air or the health. If so, the difference in market price between the two units should, *ceteris paribus,* reflect the willingness to pay for better air quality/healthy environment. Similarly, wage differentials in similar jobs can be attributed to working and living conditions. In other words, a higher wage is needed to attract workers to polluted environments or unhealthy industries like coal mining, nuclear complexes, etc.

Like in the TCM, here also people's willingness to pay for healthy environment can be arrived at by estimating a regression equation and then deriving the demand function. The functional form of the equation may be specified according to the good that is being valued. In the present context, hedonic pricing method is appropriate to derive the users' valuation of health in terms of clean air, availability of quality water, etc. Here, the functional form would be as follows:

PLif [AQi, AWi, QWi, Spi, OTi]

where

PLi   Price of the property (house)
AQi   Air quality
SQi   Soil quality
AWi   Availability of water
QWi   Quality of water
SPi   Size of the property
OTi   Vector of other attributes like distance from the market, other amenities available, neighbourhood, etc.
i   Index of properties ranging from 1 to n

One or more of these environmental qualities of the property (house) can be incorporated depending on the characteristics of the property. Hedonic pricing method is widely used with reference to air quality. The main problem associated with this method is rooted in its stringent assumption of well-functioning markets. Given the variations in property prices, this may prove useful in deriving the users' willingness to pay for various environmental qualities and health. However, interpretational problems may arise if supply adjusts to changes in prices, which may result in insignificant or zero variation in prices (of property). But, this was not found to be a frequent problem as hedonic price technique is being used effectively and satisfactorily to estimate the impact of environmental factors on property values, though it showed poor results in the context of unknown or unclear (to the individuals affected) impacts. Though these methods have the potential to deal with the valuation of natural resources, in market as well as non-market situations, one has to see whether these estimates conform to local peoples' perspective. Utmost care is needed while interpreting the results, which ought to be complemented by qualitative information.

2.6.2 Direct Methods

The most important and widely used direct method is the CVM. Of late, researchers have also employed contingent ranking or contingent behaviour to estimate individuals' willingness to pay for environmental amenities. However, the development of CVM has been very rapid due to its extensive use and hence the problems and solutions associated with it. Despite numerous criticisms levelled against it, a reasonable degree of success and persistence led to increasing attention on CVM findings.

CVM is a modern name for survey methods. Only difference is that CVM elicits how people would respond to hypothetical changes in some environmental resources. CVM deploys direct valuation questions relating to individuals willingness to pay[6] for certain environmental changes. These questions may be in the form of referendum or payment card apart from the direct questioning of the exact amount an individual/household is willing to pay (WTP). However, the direct questioning has been criticised as a difficult question to answer. The referendum approach includes dichotomous-choice, close-ended, or take-it or leave-it question formats, while the payment card format specifies a range of values from which the respondent is asked to mark the highest value he or she would be WTP. Another way of eliciting information is through a bidding game procedure, which is somewhat similar to payment card approach, where the respondent is offered different hypothetical bids until a range is generated. In this, the true willingness to pay is expected to lie between positive and negative responses rather than on a single point.

A general criticism of CVM is regarding the validity of insights derived from people's responses to hypothetical situations. How far out are the estimates

---

[6] The concept of willingness to accept is less preferred as it is observed not to reflect the true picture—often found to be giving over estimates when compared to willingness to pay estimates.

reliable or accurate? Awareness of the respondent with regard to the suggested environmental amenities, which are often esoteric (like polar bears, acid rain, rainforests, etc.), is critical for obtaining reliable estimates from CVM. Lack of knowledge regarding the 'good' or 'bad' in question may result in hypothetical answers as well. Due to this reason, CVM is the most scrutinised among the social sciences research methods, to our knowledge. However, when questions are asked regarding the issues closely related to the respondents such as health, clean water, etc., CVM is found to provide results that are more accurate. This, however, would limit the types of commodities or decisions that can be included in CVM analysis. CVM is observed to provide theoretically consistent and plausible measures of individual values for some types of environmental resources.

CVM is best suited in the event of hypothetical or missing market situations. CVM can generate reliable estimates of willingness to pay even for 'goods' or amenities in a market situation. Here, one is often talking about the improvements in the present situation rather than a status quo position. If CVM is credible in estimating non-market values, it should be at least reliable in market situations. Nevertheless, due caution has to be taken to avoid creeping in some of the important biases while using CVM. These biases include (i) sampling bias, (ii) non-response bias, (iii) strategic bias, (iv) hypothetical bias, (v) part-whole bias, (vi) information bias, (vii) aggregation bias, (viii) interviewer and respondent bias, (ix) payment vehicle bias and (x) starting point bias. Some of these biases are specific to CVM, while others are endemic to all survey methods. CVM surveys can be designed to reduce the bias problem to an acceptable level such that undertaking a CVM evaluation does provide us with useful value estimation information.

Given the poor quality of drinking water, households are WTP for improved water quality. In a study of six villages, not a single household expressed 'no' to the WTP question for the provision of quality water by a private firm (Reddy et al. 2009). This is true for both capital costs and membership fees, which is fixed at a nominal level and user charges. Majority of the households are WTP $0.75 and more as membership fee for safe drinking water. Majority of the non-poor households are WTP $1 and more (Table 2.3). In the case of user charges, all the households are WTP the present rate of $0.038 per 12 l. Most of the households prefer home delivery of water, and they are WTP extra for the transport. Among the non-poor households, 82 % are WTP extra (i.e. $0.063/12 l). In the case of

Table 2.3 Willingness to pay water (per can of 20 l) (per cent households)

Costs in $ per 12 l	Poor	Non-poor
Capital costs		
Up to $0.75	80	40
More than $1	20	60
User chargers	100	100
User charges as per cent to income	4.8	0.93
User charges as per cent to expenditure	4.5	3.5

Source Reddy et al. (2009)

poor households, only 52 % are WTP this price. The differences between poor and non-poor households in the WTP bids are due to the differences in ability to pay. Ability to pay is examined by looking at the household income and expenditure figures. As a proportion of household incomes, poor households are WTP 4.8 % against 0.93 % in the case of non-poor households. This is quite substantial by any standard, as it is often assumed that households are WTP up to 3 % of their income. In terms of expenditure, poor households are WTP more than their counterparts (Table 2.3). It is also observed that willingness to pay for public provision of drinking water is often lower when compared to private supplies. This is mainly due to lack of trust in public utilities in the developing countries.

It is observed that the willingness to pay estimates is often on the higher side when compared to HC approach. A caparison of these two methods in the Chongqing region of China revealed that WTP estimates are more than three times higher than that of the estimates from HC approach. A major share of this goes to the avoidance of premature cancer deaths due to water pollution (Yongguan et al. 2001). The difference in estimates could be due to the reason that WTP estimates often include non-tangible/non-economic costs, viz. social, psychological, aesthetic values, etc.

Contingent ranking or contingent behaviour is an indirect approach within the direct methods. In contingent ranking, respondents are asked to rank some non-market resources in the order of their preference. These non-monetary preferences are then 'anchored' by simultaneously asking respondents about some of the familiar items like hand pump or bore well. Then, respondents are asked for their willingness to pay (WTP) for the familiar items. These WTP estimates are then used to infer the WTP for non-market sources. Though this method appears to be simple and capable of generating reliable WTP estimates, it may not provide the true estimates of WTP apart from having additional biases, which are not identified so far. Another pertinent problem may arise due to the difference in ranking non-monetary priorities of individuals and their valuation in monetary terms. This mismatching of monetary and non-monetary priorities is found to be substantial in the context of local-level valuation of resources.

In the case of contingent behaviour approach, respondents are provided with alternative scenarios of environmental amenities from which they have to make a choice. This facilitates the explanation for the choice of one alternative over others as a function of attributes, which include travel distance, etc. This method makes data generation more complex, as it tends to confuse respondents by giving them multiple choice of amenities. This method also involves more sophisticated econometric techniques for estimation purposes. More importantly, this method when combined (jointly estimated) with revealed preference (indirect) approach is expected to provide a fruitful variant to CVM rather than by itself.

Keywords and Definitions

AC	Averting costs
CBA	Cost–benefit analysis
CVM	Contingent valuation method
DA	Decision analysis
HC	Human capital
Life-cycle	'Consecutive and interlinked stages of a product system, from raw material acquisition or generation from natural resources to final disposal' (ISO 2006)
Life-cycle approaches	'Techniques and tools to inventory and assess the impacts along the life cycle of products'
Life-cycle assessment (LCA)	'Compilation and evaluation of the inputs, outputs and the potential environmental impacts of a product system throughout life-cycle costing' (LCC): 'Life-cycle costing, or LCC, is a compilation and assessment of all costs related to a product, over its entire life cycle, from production to use, maintenance and disposal' (UNEP/SETAC 2009)
MCA	Multi-criteria analysis
RBA	Risk–benefit analysis
TCM	Travel cost method
WTP	Willingness to pay

References

Barringer HP (2003) A life cycle cost summary. In: International conference of maintenance societies (ICOMS®-2003, Presented by Maintenance Engineering Society of Australia), May 20–23, Sheraton Hotel Perth, Western Australia, Australia

Barringer HP, Weber DP (1996) Life cycle cost tutorial. In: Fifth international conference on process plant reliability (Organized by Gulf Publishing Company and Hydrocarbon Processing), Oct 2–4 (Revised Dec 2), Marriott Houston Westside Houston, Texas

Finnveden G, Hauschild MZ, Ekvall T, Guinée J, Heijungs R, Hellweg S, Koehler A, Pennington D, Suh S (2009) Recent developments in life cycle assessment. J Environ Manage 91:1–21

Guinée JB, Gorrée M, Heijungs R, Huppes G, Kleijn R, de Koning A, van Oers L, Wegener Sleeswijk A, Suh S, Udo de Haes HA, de Bruijn H, van Duin R, Huijbregts MAJ (2002) Handbook on life cycleassessment. Operational guide to the ISO standards. I: LCA in perspective. IIa: Guide. IIb: Operational annex.III: Scientific background. Kluwer Academic Publishers, ISBN 1-4020-0228-9, Dordrecht, pp 692 doi: 10.1007/978-3-319-06287-7_2

Hoff H (2011) Understanding the nexus. Background paper for the Bonn 2011 conference: the water, energy and food security Nexus. Stockholm Environment Institute, Stockholm

ISO 14040 (2006) Environmental management-life-cycle assessment-principles and framework, Geneva, Switzerland

Kurian M, Ardakanian R (2013) Institutional arrangements and governance structures that advance the nexus approach to management of environmental resources. Paper prepared for draft white book: advancing a nexus approach to the sustainable management of water, soil and waste. International kick-off workshop on 11–12 Nov 2013, UNU-FLORES. Retrieved 10 Apr 2014. From http://flores.unu.edu/wp-content/uploads/2013/08/FINAL_WEB_whitebook.pdf

Kurian M, Reddy VR, Rao RM, Lata S (2008). Adapting to climate variability: productive use of domestic wastewater as a risk management option in peri-urban regions. Research report. Water and Sanitation Programme, The World Bank, New Delhi

Lundin M (2002) Indicators for measuring the sustainability of urban water systems: a life cycle approach. Environmental systems analysis. Chalmers University of Technology, Canada

McConville JR (2006) Applying life cycle thinking to international water and sanitation development projects: an assessment tool for project managers in sustainable development work. A report submitted in partial fulfilment of the requirements for the degree of Master of Science in Environmental Engineering, Michigan Technological University

Reddy VR (1999) Quenching the thirst: the cost of water in fragile environments. Dev Change 30:79–113

Reddy VR, Behera B (2006) Impact of water pollution on rural communities: an economic analysis. Ecol Econ 58(3):520–537

Reddy VR, Batchelor C, Snehalatha M, Rama Mohan Rao MS, Venkataswamy M, Ramachandrudu MV (2009) Costs of providing sustainable water, sanitation and hygiene services in rural and peri-urban India. WASH Cost-CESS working paper no. 1. Centre for Economic and Social Studies, Hyderabad

Reddy VR, Kurian M (2010) Approaches to economic and environmental valuation of domestic wastewater. In: Kurian M, Patricia M (eds) Peri-urban Water and Sanitation Services: Policy, Planning and Method, Springer, London.

UNEP/SETAC (2009) Guidelines for social life cycle assessment of products. Paris

Yongguan C, Seip HM, Vennemo H (2001) The environmental costs of water pollution in Chongqing, China. Environ Dev Econ 6(3):313–333

Chapter 3
LCCA Applications in Infrastructure and Other Projects: Some Case Studies

3.1 LCCA Application in Real World

Application of LCCA in the real world is far from perfect. Often, its application is limited to specific phases of the life cycle and does not include the environmental aspects. In the case of infrastructure projects, only positive environmental impacts are valued and incorporated as benefits. It is rare to find an all-inclusive study that combines all the cost parameters, energy use and implication on the life-cycle performance. Project selection, however, requires an integrated evaluation of an energy-saving infrastructure. For instance, the life-cycle analysis of energy-efficient buildings accounts for energy savings when compared with other alternatives. At the same time, it may not include the benefits from the savings or mitigating the impacts of greenhouse gases. In most cases, LCCA is adopted in infrastructure projects mainly to identify and chose the least cost or cost-effective project from the available alternatives.

The least cost options are not necessarily environmentally sustainable options, particularly since green technologies can be more expensive, at least initially. In the context of applying LCCA to natural resource management, the externality aspects are being incorporated of late. These include externalities associated with water savings, use of chemical fertilisers and manure. Often these estimates are based on assumptions that are not location specific. The studies limit themselves to estimating the magnitude of such externalities based on assumptions rather than measuring and valuing them for each case. This provides the second best assessment of the environmental impacts associated with the natural resource management.

In the case of infrastructure, projects that are associated with service delivery do not integrate environmental impacts and their related externalities. The studies dealing with public services like water and sanitation appears to be more effective when they tend to assess the changes in lifespan of the projects. At the same time,

© The Author(s) 2015
V.R. Reddy et al., *Life-cycle Cost Approach for Management of Environmental Resources*, SpringerBriefs in Environmental Science, DOI 10.1007/978-3-319-06287-7_3

they limit themselves to making comparative assessments of different technological options. Environmental impacts of these technologies are often not integrated into these assessments.

3.1.1 Data Requirements and Limitations

Most of the studies adopt the standard life-cycle costing concepts and components. The cost components include infrastructure costs, replacement costs and maintenance costs. The detailed composition of these cost components makes it difficult to get data, as some of the cost components are neither available nor allocated in the real world. Hence, these costs need to be incorporated by using assumptions or some kind of guesstimates while evaluating the new projects. In the case of post-evaluation, these costs need to be culled from the old records, which is cumbersome. Besides, the old data may not be in the desired format. Nevertheless, post-assessment provides insights into how cost components could be assessed in the new projects (Reddy et al. 2012).

Thus, availability of data in an appropriate form is a prerequisite for the adoption of LCCA in a comprehensive manner. The main challenge is obtaining the data pertaining to environmental aspects at the national, regional and sub-regional level. Such data are not available in the developing countries. Initiating the data collection process needs to be treated as a priority. The challenge is to identify various indicators of environmental impacts associated with various sectors. The comprehensive and complex nature of information makes the collection not only difficult but also costly. A consultative process is necessary for establishing the most appropriate database for adopting LCCA in the specific country. Moreover, some countries may need financial support to initiate and set up a data collating agency or data bank.

3.1.2 Profile of the Projects

Typically, the infrastructure projects where LCCA is adopted include construction, roads and power sector. A comparative assessment of different methods of construction, materials used, technologies adopted, etc. is carried out in order to identify best options. Adoption of LCCA in infrastructure projects is limited to developing countries. The assessments are carried out mainly to identify the cost-effective alternatives prior to investment decisions. In most cases, the assessments are limited to the cost components of the life cycle and do not include the environmental impacts, and hence, these could be categorised as LCA studies rather than LCCA studies.

The natural resource-based projects where LCCA is adopted include mainly water, crops and biofuels. Of these, water-related projects focus on sustainable service delivery in the drinking water sector. Per unit costs of different technologies are assessed in a post-project approach. These provide information on actual costs for various components in realistic manner. At the same time, these assessments are limited to LCA and do not include the environmental impacts associated with drinking water. Crop-related projects are more comprehensive in adopting LCCA. Contrary to infrastructure projects, the main aim of crop-related assessments is to identify and assess the environmental impacts. Interestingly, most of these studies are from developing countries. In the following section, some important and relevant case studies are presented.

3.2 Case Studies

3.2.1 Infrastructure Projects (Construction, Roads, etc.)

Case Study (a): *Kneifel, Joshua (March* 2010). *Life-cycle carbon and cost analysis of energy efficiency measures in new commercial buildings. Energy and Buildings*, 42(3): 333–340.

In this case study, a prototypical building is used for a comparative assessment of three design alternatives, viz. using different levels of energy efficiency standards. The energy efficiency designs vary in terms of insulation material, windows, thickness of the walls and roof decks. It is expected that the low-energy design would save up to 25–30 % energy when compared to earlier standards.

3.2.1.1 Approach

A whole-building design approach is adopted including life-cycle assessment along with energy simulations of the design. Cost implications are derived using the life-cycle cost analysis. Cost components include building construction costs for various works, including the contractor and architect fee; maintenance, repair and replacement costs.[1] These costs are taken from the reference manual. In the case of new components, repair costs are assumed 1 percentage of the costs. Energy costs are estimated using the energy consumption data and fuel and natural gas rates or the retail price of energy.

Using this cost information, the study has computed life-cycle costs for three different building designs. While the ASHRAE 20.1–2004 is considered the base case, the two alternatives are ASHRAE 20.1–2007 and the low-energy case (LEC) designs. The cost difference between the base case and the alternatives provides the net savings. Moreover, the internal rate of return is estimated for all three cases. The higher the difference between internal rate of return and market rate of return (interest rate), the most preferred the energy-saving design.[2]

---

[1] Whitestone Research and Building Maintenance and Repair Cost Reference are used for these costs.

[2] NIST's BEES software is used to compute the life-cycle costs for the building design alternatives.

3.2.1.2 Results

The results indicate that LCCA provides a comprehensive cost estimate, as operating costs are often not included. LCCA also provides a strong rationale for making appropriate investment decisions that is sustainable in the end. The comparative cost assessments show that energy-efficient buildings tend to be cost-effective as the life of the building increases for the energy-savings measures are capable of reducing carbon emissions by 32 % over a period of ten years. This calls for a systematic and all-inclusive approach looking at various available design options.

Case Study (b): *Krützfeldt, Gerard.* (September 2012). *Life-cycle costing and risk management: the influence of uncertainties on Dutch transportation infrastructure projects. MSc Thesis Report for Construction Management and Engineering, Faculty of Civil Engineering and Geosciences, Delft University of Technology.*

This case study presents alternative transport options in a specific location in the Netherlands in order to reduce traffic congestion and improve quality of life. The LCCA performed in this case incorporates the risk-based analysis pertaining to consumer delays and the related social impacts. The assessment was carried out for three alternatives, viz. a fully immersed tunnel, a bridge and a semi-underground tunnel with the assumption that the benefits from these alternatives are the same.

3.2.1.3 Cost Components

Investment costs and maintenance costs are the broad cost components. Investment costs include construction costs, property costs, engineering costs, other additional costs and costs associated with risk and uncertainties. Maintenance costs include asphalt, tunnel technical installations (TTIs), concrete work, other additional costs and risk and uncertainty related costs. It is assumed that construction and additional costs will take place between 2014 and 2016, and engineering costs between 2012 and 2016. Property costs are taken at 2012 to 2015 rates. Quantities and unit costs are estimated using cost data and expert opinions. Construction is expected to be completed by 2017.

For estimating the net present values (NPVs), an economic life of 30 years is assumed and a discount rate of 2.5 %, which could go up to 10 % (as per the contractors /market rate). It is assumed that major replacement will take place only during year 30. NPV is calculated as of 1 January 2012. Value added tax (VAT) of 19 % is included in the total cost. Straight-line depreciation over lifetime is used, and residual costs are assumed zero at the end of technical lifetime (100 years).

Including user delay (traffic delays) costs in the LCCA serves as a basis for comparison between alternatives. Different scenarios are built around this using various assumptions such as state of the economy (strong/weak), value of time and estimation of number of persons per vehicle. Since data are not available on future indicators, broad assumptions are made using present trends. Similarly, some risks are expected due to fire, closure for reconstruction activities affecting user delay costs.[3]

3.2.1.4 Analysis

It is observed that LCC provides reasonable results after incorporating user delay costs (level II); cost of failure and additional future risks using probabilistic estimations (level III). At the same time in the event of uncertainties, it would be difficult to make rational decisions within the context of social, political and/or environmental issues. However, the level of details should be good enough to decide on maintenance strategies, which is prerequisite for calling bids. Though user delay costs could be excluded when their share is marginal in the total costs, it is argued that inclusion of these costs could yield interesting opportunities to improve design, develop maintenance strategies, and more important, to provide insights into maintenance contracts, possibly an indirect benefit of LCCA.

It is observed that discount rates could influence LCCA estimates substantially. A low discount rate could lower annual expenses and enhance the viability of the project. Lower discount rates are preferred in the case of environmental and social projects. And higher discount rates would result in the opposite. Often different discount rates are used for different types of industries, markets, commodities, etc. Interest and net inflation rates can also be used as discount rates. Using a sensitivity analyses on different rates (ranging from low to high) is preferred prior to investment decisions.

Life of the project has major influence on costs. Number of factors influences the lifespan of the projects or systems. These include physical stress/load, chemical degradation, environmental effects (natural disasters), human behaviour, legislation effects, politics and other risks. High unpredictability of some of these factors not only makes assessment of economic lifetime difficult but also makes the economic lifetime shorter than the expected or normative lifespan. For instance, functional capacity is changing faster than technical requirements (economic growth is changing faster than rising sea levels). For the purpose of LCC, either technical life (50, 80 or 100 years) or an economic life can be used. Both can be used separately to perform sensitivity analysis.

3.2.1.5 Results and Discussion

LCC can serve as an effective tool that looks at not only benefits, but also considers environmental and social effects, deterioration factors, obsolescence drivers and political contexts in addition to costs. However, it may be noted that LCCA is not a silver bullet or an all-encompassing problem-solving technique or tool. A primary shortcoming is the sensitivity analysis where equal weight is

---

[3] Probability of failure obtained from Infrastructure Department.

accorded to all cases irrespective of the probability of occurrence. Though this is not realistic, they can provide comparative insights.

The comparative analyses of three alternatives in the present case, alternative 2, the bridge, is the most cost-effective one in the context of a life-cycle cost analysis. Alternative 2 should be the preferred option irrespective of the discount rate used. Interest rates and maintenance tend to influence results heavily in the case of the tunnel. Alternative 2 (bridge) is found to be the most cost-effective in both investment and maintenance costs. And, its NPV is lower than all the others are. Cost per year analysis indicates that interest rates account for much more regarding fixed costs of the bridge. Maintenance still accounts for a large percentage, so it should still be regarded, as there are some cost items in maintenance activities that incur large costs.

When uncertainties pertaining to cost and quantity using the probabilistic estimation method (level III) are included, the technical installations provide the most critical (both in terms of costs and quantities), followed by their replacement, etc. The tunnel technical installations (TTIs) should be the main focus of attention as they contribute the most to the LCC and therefore explain high costs in maintenance activities. Increase in cost of maintenance activities is observed due to increase in labour costs, rise in oil prices having an impact on asphalt and others. The largest cost contributors are maintenance, interest and depreciation. This is due to the above-mentioned TTIs and other expensive items in maintenance activities.

It is observed that reducing the life-cycle analysis can increase level of confidence and the Monte Carlo analysis should be done with a fixed risk-free discount rate. This means the NPV remains unchanged. When looking at discounted cash flows for calculation of NPV, the residual value is subject to the period of analysis. In practice, public infrastructure projects have no resale value, unlike other types of private organisations or industries where the concept of salvage value has more importance. Interest rates account for a larger share in the fixed costs of the bridge. Maintenance still accounts for a larger share in the total cost and hence needs attention. However, the higher the interest rate, the less effect other costs will have on total fixed costs. For example, an interest rate of 8 percentage constitutes more than 50 % of the total fixed costs for both cases, whereas the other costs now contribute less to the total percentage.

3.2.2 Natural Resource-Based Projects (Drinking Water and Sanitation (WASH), Crop Systems, Bioethanol, etc.)

Case Study (a): *Iraldo, Fabio, Testa, Francesco, Bartolozzi, Irene* (2014). *An application of Life Cycle Assessment (LCA) as a green marketing tool for agricultural products: the case of extra-virgin olive oil in Val di Cornia, Italy. Journal of Environmental Planning and Management,* 57(1): 78–103. doi:10.1080/09640568.2012.735991.

This case study assesses the potential environmental impacts of Val di Cornia extra-virgin olive oil and of its production chain using LCCA with an aim to combine eco-friendly production processes and competitive advantages for local producers. In

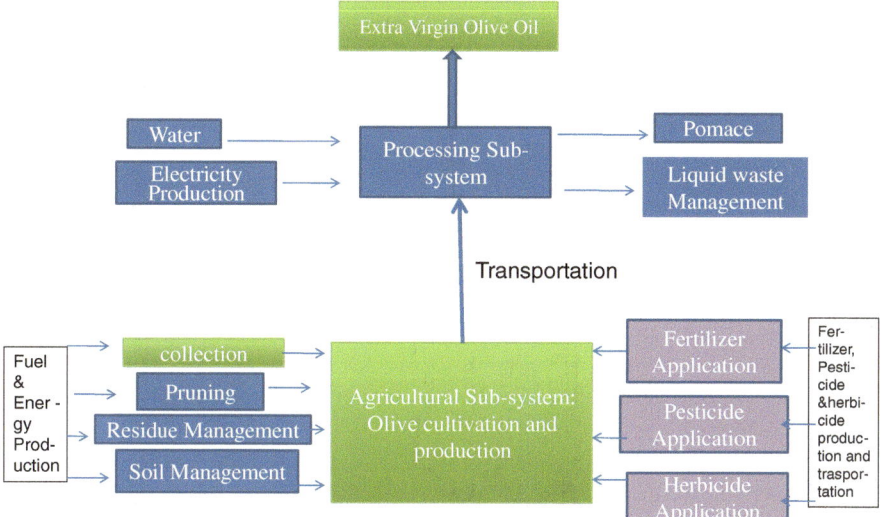

Fig. 3.1 Flow chart of the processes for olive oil production (system boundaries). *Source* Iraldo et al. (2014)

addition, the case provides a pilot experience for a local environmental qualification scheme to support local policies for sustainable production and consumption patterns.

3.2.2.1 System Boundaries

Since the focus was on the environmental impacts of the production cycle, the analysis is restricted to a 'cradle-to-gate' analysis and the system boundaries include the processes within the agricultural and olive processing stages (Fig. 3.1). In the farming stage, all crop activities from ploughing, application of fertilisers, pesticides, harvesting and transporting are included. The externalities associated with the production process are taken into account. These include production and transportation of the chemical inputs, water, fuel and energy. The diffused emissions due to fertilisers and pesticides were calculated. The carbon balance of the olive groves was also included. Pruning residues are either burned for home heating or shredded and spread on the fields. Solid waste produced in this stage consists mainly of the plastic packaging of the chemicals used. The olive milling process involves energy and water consumption. Finally, extra-virgin olive oil and pomace are produced and sold to a treatment factory and processed for energy use. The olive mill wastewaters are usually treated as liquid waste.

3.2.2.2 Approach

The study is based on a sample of seven olive growers covering 2.2 % of the total area under olive crop. A structured questionnaire was used to collect the information on crop production. Estimates of emissions are arrived at using the secondary sources. Non-renewable fossil energy consumption was evaluated using cumulative energy demand method. A kg extra-virgin olive oil is used as a functional unit. All the inputs and emissions are standardised per kg of extra-virgin olive oil. These include water consumption, energy consumed, waste produced and emissions, at the production and processing stages. Since extra-virgin olive oil and pomace are coproducts, environmental impacts were accordingly allocated economically.

3.2.2.3 Results and Interpretation

Impacts are observed in all categories at the agricultural production stage of olive oil. Chemical inputs or components (pesticides) are the main contributors through acidification, eutrophication and water consumption impact categories. Fuel consumption in the farming operations also contributes to the degradation of non-renewable energy sources. It is argued that environmental impacts could be reduced through cutting down on the use of pesticides and fertilisers. Similarly, at the processing stage, use of wastewater as fertiliser could provide significant environmental benefits. The sensitivity analysis suggests that the scenario where reuse of pruning residues and wastewater from processing as fertilisers provides the maximum benefits when compared to other scenarios where neither of the practices are followed (worst case) and either of the practices is followed (second best).

Case Study (B): *Sawaengsak, Wanchat, Silalertruksa, Thapat, Bangviwat, Athikom, and Gheewala, Shabbir H* (2014) *Life cycle cost of biodiesel production from microalgae in Thailand. Energy for Sustainable Development,* 18: 67–74.

This case study by adopting LCCA evaluates the economic feasibility of biodiesel production from microalgal oil and other high-value chemicals in both open ponds and closed photo-bioreactors in Thailand. Biomass-based liquid fuels such as biodiesel and bioethanol are good substitutes for petroleum-based fuels. While there is competition for land between food and biomass production in arable lands, degraded lands open up avenues for exploring the potential for biomass-based fuel production. Microalgae are among the potential sources of biodiesel though it is used to produce nutritional food, viz. algal meal, omega-3 and fatty acids, in countries such as Thailand. One important issue in this regard is achieving least cost options in terms of capital and operational costs.

3.2.2.4 Materials and Methods

The study site is a commercial algal facility located in Chiang Mai province in the northern region of Thailand. The study provides a comparative picture of open pond and photo-bioreactor algae farms. The lifespan of the plant equipment is assumed to be 15 years with a production target of 720,000 l per year (assumption).

Additionally, using microalgal residue as aquaculture feed is also considered by comparing anaerobic digestion in terms of annual profitability. The process used to refine the oil and produce biodiesel is assumed the same as that used for the production of biodiesel from palm oil in Thailand. The algal oil is transported for biodiesel production over a distance of 32 km.

The following four scenarios were examined: (1) a base case of pond system (without omega-3 fatty acid production), (2) an alternative case of pond system (with omega-3 fatty acid production), (3) a base case of photo-bioreactor system (without omega-3 fatty acid production) and (4) an alternative case of photo-bioreactor system (with omega-3 fatty acid production). The base case of two systems is projected to produce 720,000 l of biodiesel per year along with glycerine at 62.2 tons per year as a by-product. The alternative case is projected to produce 432,053 l per year of biodiesel plus 239.02 tons of omega-3 fatty acids and 37.3 tons of glycerine per year.

Capital and operating costs are obtained from secondary sources. The costs are converted to Thai Bahts and adjusted to 2012 prices using the annual inflation rates. Maintenance costs are assumed at 2 percentage of the capital cost, interest rate at 7.31, 30 % as contingency, depreciation at 6.7 % and tax rate is 20 %. Capital is borrowed for a ten-year period.

3.2.2.5 Results and Discussion

It is assessed that of the two options, photo-bioreactors have an advantage over raceway ponds with respect to land and water requirements. On the other hand, electricity required for photo-bioreactors is significantly greater than raceway ponds. Capital costs are higher than operating costs for both the alternatives. Despite the fact that alternative case can obtain more income from the sale of by-products, it is not enough to make it profitable. Due to high annual operating expenses, the cost of biodiesel production is substantially higher compared to the current biodiesel market price. However, the net present values turned out to be negative for both the alternatives. It shows that the commercial algal biodiesel production is not profitable even after selling omega-3 fatty acids. Financial feasibility of photo-bioreactor is substantially lower than raceway ponds due to the higher capital investment as well as much higher operating cost from electricity.

Given the difference between market price of biodiesel (28.8 THB/l) and algal biodiesel (68 THB/l), the latter is not economical. Only by reducing the costs by 50 % along with tax exemptions would the production of algal biodiesel be economically viable. Production costs could be reduced by increasing production through adoption of new technologies or improvement of productivity of algal biomass. Positive net present values with reasonable rates of return are only possible with moderately high yields (134 kg/ha) and lipid concentration coupled with higher crude oil price and subsidies.

Case Study (c): *Gathorne-Hardy, Alfred* (2013) *A Life Cycle Assessment of Four Rice Production Systems: High Yielding Varieties, Rainfed Rice, System of*

Rice Intensification and Organic Rice. International Symposium on Technology, Jobs and a Lower Carbon Future: Methods, Substance and Ideas for the Informal Economy (The case of rice in India); India International Centre (Annexe), 40, Max Mueller Marg, New Delhi, 13-14, June 2013.

Agriculture is directly responsible for about 10–12 % of global greenhouse gas (GHG) emissions and indirectly for roughly another 10 %. This case study uses LCCA to assess the water use and the energy requirements of four different rice production technologies—intensive flooded high-yielding varieties (HYV), rainfed rice, systems of rice intensification (SRI), and organic rice.

3.2.2.6 Approach

Survey methods were used to collect data from three different locations in the semi-arid regions of South and East India covering the four rice-growing methods. System boundaries and functional unit are defined (Fig. 3.2). One kg of paddy at the farm gate is set as a functional unit. System boundaries (within the red line) include the entire production process including embodied energy and emissions. Methane emissions from livestock and indirect land use changes are optional. On the other hand, embodied water used for energy and machinery, embodied water

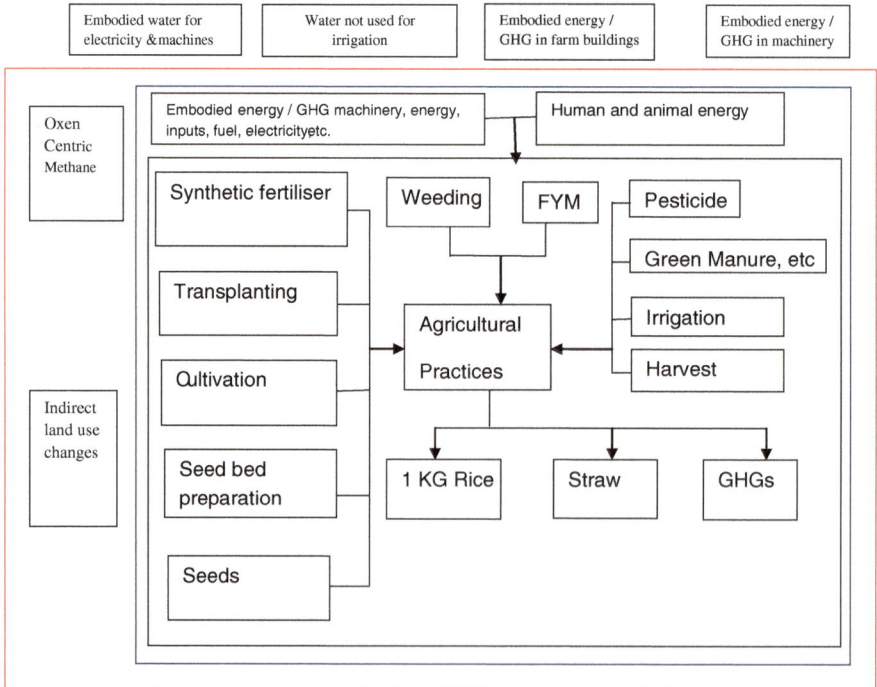

Fig. 3.2 System boundaries for rice. *Source* Adopted from Gathorne-Hardy (2013a), b

not used for irrigation, embodied energy/GHG from farm buildings and machinery are not included in the system boundaries.[4]

3.2.2.7 Results and Discussion

There is no significant difference in the GHG emissions associated with paddy production between the different production systems investigated: all produce approximately 1 kg CO_2-eq for each kg of paddy. Water management and reducing the demand for water reduces embodied GHG and energy demands as well as potentially reducing groundwater-based GHG emissions. It is observed that switching to System of Rice Intensification (SRI) method could reduce CH_4 emissions if excessive FYM is not applied. While converting to SRI would reduce water requirements per kg of paddy, there is no significant reduction in the total water extracted per hectare. Improving transmission and distribution efficiencies would produce substantial savings in GHG emissions. Emissions from organic rice are equivalent to that of HYV in most instances. Organic production showed the highest nitrogen-use efficiency compared to all other systems. There is a chance that in order to maintain yields, additional FYM inputs would be needed, with the inevitable CH_4 emissions. It is concluded that adopting SRI to organic systems would yield further gains, with significant energy savings and environmental benefits from reduced pesticides and fertiliser use.

Case Study (d): *Reddy, V. Ratna, Venkataswamy, M. and Snehalatha, M (2013) Unit costs and service levels: Technology wise. In Sustainable water and sanitation services: The life-cycle cost approach to planning and management. London: Routledge, Earthscan.*

This case study compares the normative allocations with the actual expenditure across technologies used in Andhra Pradesh, India. The actual costs are estimated based on the data collected from 187 habitations spread over the nine agroclimatic regions in the state for various technologies. The objectives include cost of service provision per technology between actual and normative lifespan and relative expenditure on different cost components in reality against the agency norms. Environmental impacts not included.

3.2.2.8 Approach

The life-cycle cost (LCC) approach is adopted to estimate the actual cost components of service provision. The costs assessed here cover the construction and maintenance of systems in the short and long term, taking into account the need for hardware and software, operation and maintenance, cost of capital, source protection, and the need for direct and indirect support costs, including

---

[4] IPCC (2007) 100-year global warming potentials were used to calculate GHG equivalents.

training, planning. Cost data were obtained from the official records of the RWSS Department at the district level. These data were triangulated or cross-checked with the help of the data generated from the village panchayat (local government). The data on operation and maintenance were obtained from the village panchayat records.

Capital expenditure has two components, namely hardware (CapExHrd) and software (CapExSft). Establishment of water infrastructure, water extracting elements, purification equipment, storage reservoirs, distribution systems, etc., are part of capital expenditure on hardware, while capital expenditure on software includes the costs for planning and designing the water schemes at the village level. The capital costs, hardware as well as software, are one-time costs.

For the purpose of the present analysis, we have considered only investments in infrastructure that are still functional. In most cases, the system or infrastructure is non-functional when the source fails beyond rehabilitation, for example, drying up or collapse of a bore well. All the capital investments are cumulated over the years. Capital maintenance expenditure (CapManEx) is another major expenditure for the renewal and rehabilitation of the systems, i.e. replacement of major equipment such as pump sets, bore holes, plant equipment and distribution systems. Capital maintenance expenditure is also summed up over the years.

Operational expenditure (OpEx) for the regular maintenance of the systems is incurred annually and hence considered as the average of the years for which data are available after bringing them to the current year. Expenditure on direct support costs (ExDS) is in the form of salaries to the staff, IEC activities, demand management initiatives, etc., while expenditure on indirect support costs (ExIDS) is the costs associated with macroplanning and policymaking at the national and state levels. These costs are estimated based on the data from the planning and budgetary documents with the help of some assumptions and expert opinions.

Since capital and capital maintenance expenditure are one-time investments, in the past they were converted to current values (2010) using the National GDP inflator for the specific years and converted to US dollars using the average 2010 exchange rate. These costs are annualised using the normative lifespan and observed lifespan of the systems. The data on normative lifespan are provided by the department, which is nothing but the expected lifespan of a specific component. The observed lifespan is the actual number of years the system (major component) lasts.

In the case of departmental cost figures, the latest (2010) estimates for different systems are considered. Estimates are provided for single and multi-village schemes separately. Since the actual costs include both these sources, in most cases, the average of both is taken. The official cost estimates do not include the salary component of the direct support costs (ExDS) and the indirect support costs (ExIDS). These two components, which are estimated using budget data, are added to the official norms in order to make them comparable with the actual costs based on our estimates.

The existing technologies prevalent in the rural water supply include hand pumps (HPs), direct pumping (DP) or mini-piped water supply (MPWS), single-village schemes (SVS) and multi-village schemes (MVS). These technologies are used in 107 of the 187 sample villages, while the remaining sample villages use different combinations of these four technologies.

3.2.2.9 Results and Discussion

Fixed costs include the capital expenditure on hardware (infrastructure) and software (planning and design). When the per capita sum of (cumulative) capital costs is taken into account, multi-village schemes are, relatively speaking, the most expensive of the pure technologies. Hand pumps are the cheapest, followed by DP/MPWS and SVS. Per capita costs are more in the case of villages that are served by multiple schemes. However, cost differences are statistically significant only in the case of hand pumps and the combination of MPWS + SVS + MVS. That is, the per capita costs of hand pumps are significantly cheaper, while those for MPWS + SVS + MVS are significantly higher when compared to the other technologies. The differences in the per capita costs among the pure technologies (MPWS, SVS and MVS) are not significantly different (Fig. 3.3).

Annualised unit costs were calculated for normative as well as observed life of the schemes. While the normative unit costs reflect the ideal conditions of good asset management, observed unit costs represent the actual picture in the present management system. The normative lifespan is worked out on the basis of economic and useful life of the systems, while the observed lifespan is the life of the systems in reality. The normative lifespan data are provided by the department (RWSS). Future service delivery requirements and their cost norms are arrived at by the department on the basis of the normative lifespan of the systems.

The observed lifespan is often found to be lower because the systems breakdown frequently due to lack of maintenance or due to the hydrogeology of the region (bore well failure). Moreover, poor design and implementation also speed up the decay of the systems. Similarly, in the case of new systems where breakdowns are few, the observed lifespan could be lower, pushing the costs up.

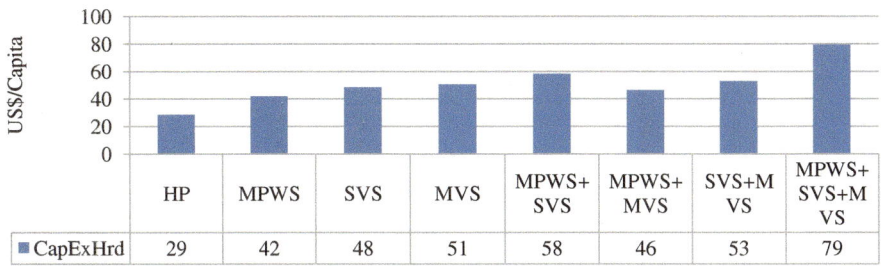

Fig. 3.3 Capital (cumulative) expenditure per capita across technologies

3.2.2.10 Recurrent Costs (CapManEx, OpEx, ExIDS, ExDS)

As far as pure technologies are concerned, the recurring costs range between US$0.9 per capita per year in the case of HPs and US$2.8 in the case of MPWS (Fig. 3.2). Recurring costs of single and multi-village schemes are same, at US$2.5 per capita per year. The unit costs are as high as US$6.3 per capita per year in the case of a combination of three technologies. These cost differences are marginal and significant statistically only in the case of HPs and the combination of three technologies (MPWS + SVS + MVS).

Hand pumps are the cheapest even in terms of recurring costs. The high unit costs in the villages that are served with a combination of technologies could be due to the multiple schemes that need maintenance. This is because all the technologies are functional and hence incur operation and maintenance costs along with other recurring costs. In some cases, the villages are upgraded to multi-village schemes due to political reasons though they may not be in need of improved service levels; and the unit costs of the combination of three technologies (MPWS + SVS + MVS) are significantly higher than that of the other technologies. The observed differences in unit costs are not, however, significantly different between any of the pure technologies (Fig. 3.4).

3.2.2.11 Unit Cost Versus Service Provided Per Technology

More than 50 % of the households receive basic and above service in terms of quantity, quality and reliability (Fig. 3.5). Accessibility, measured in terms of the time spent on fetching water, appears to be a major concern irrespective of the technology used. At the highest level, only 36 % of the households spend less than 30 min a day fetching water (receiving above basic service) in the case of villages that have three technologies functioning simultaneously (MPWS + SVS + MVS).

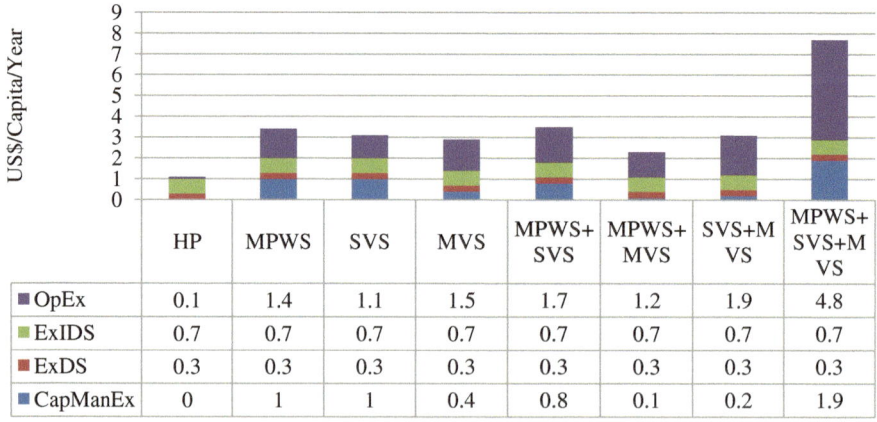

US$/Capita/Year	HP	MPWS	SVS	MVS	MPWS+ SVS	MPWS+ MVS	SVS+M VS	MPWS+ SVS+M VS
OpEx	0.1	1.4	1.1	1.5	1.7	1.2	1.9	4.8
ExIDS	0.7	0.7	0.7	0.7	0.7	0.7	0.7	0.7
ExDS	0.3	0.3	0.3	0.3	0.3	0.3	0.3	0.3
CapManEx	0	1	1	0.4	0.8	0.1	0.2	1.9

Fig. 3.4 Average cost of provision

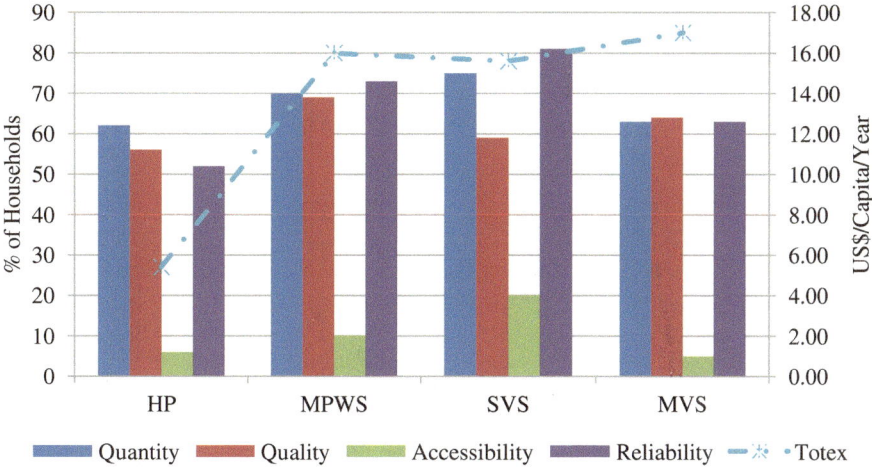

Fig. 3.5 Service levels (basic and above) and unit cost across technologies

Accessibility is the lowest among MVS and HP villages. Unit costs would go up in the case of low accessibility when opportunity costs of time are taken into account. Service levels are poor for all indicators in the case of HP villages. Among the pure technologies, SVS is providing better service in terms of all the indicators except quality followed by MPWS, while MVS villages compare poorly with both these technologies in all the indicators except quality, which is slightly better than that of SVS. The centralised distribution systems of MVS do not seem to be efficient in service delivery, and the differences in service levels are not statistically significant in most cases, especially among the pure technologies.

However, as mentioned earlier, the differences in unit costs are not very different among these technologies, except that HP is the cheapest and the combination of three technologies (MPWS + SVS + MVS) is the most expensive. When unit costs are plotted against the service levels, it is clear that while HP is associated with poor service levels, the most expensive technology provides only marginally better service in the case of quantity, quality and accessibility. On the other hand, SVS and MPWS provide relatively better services when compared to MVS. It may be noted that better quality and accessibility is also associated with buying water. In the absence of buying water, MVS would do well in terms of quality, due to its dependence on surface water sources. While these service levels are based on the proportion of households receiving a specific level of service, the actual cost of provision in terms of cost per unit of water is not captured here. Since the cost per unit of water is an important indicator while comparing the technologies, this aspect is covered in the following section.

3.2.2.12 Cost Per Unit of Water

Total water consumption for the year under each technology is compared with the annualised cost per capita for the specific technology. The ratio between the annualised cost and the annual water use of the habitation gives the cost per unit of water. While the per capita service level is not very different across the technologies, especially the pure technologies, the cost per unit of water, at the aggregate level, varies (Table 3.1).

The cost per unit of water is the lowest in the case of hand pumps and highest in the case of MVS as far as pure technologies are concerned in terms of normative as well as observed lifespan of the systems. While costs in HP-dependent villages are low, their service levels are also low, especially in terms of reliability and accessibility. Single-village schemes appear to be the best of the lot with better service indicators and relatively low costs, in terms of the cost per capita per year as well as the cost per unit of water. On the other hand, MVS has relatively higher unit costs with low service levels when compared to MPWS.

The cost estimates using the life-cycle costs approach (LCCA) bring out the following important issues: Across the technologies, the average unit costs are about three times lower for hand pumps. Multi-village schemes are relatively more expensive though the cost differences are not statistically significant. Multi-village schemes are associated with high (cumulative) capital costs with wide variations. Cost composition as well as their shares varies across locations. Cost composition is presently focused on infrastructure to the neglect of other important components such as source protection, capital maintenance and quality. All the technologies are associated with high recurring costs when compared to hand pumps, especially the operation and maintenance costs. On the other hand, capital maintenance costs are more in the case of SVS and MPWS schemes. As far as service levels are concerned, hand pumps provide poor services in terms of reliability, accessibility and quality. Moreover, hand pumps are not the commonly used technology, as they are used mostly to cope with scarcity conditions. At the policy level also, it is not a policy option due to the low preference at the community level. Single-village schemes perform better in the case of service levels in terms of all four indicators. Multi-village schemes are expensive even in terms of cost per unit of water despite their larger coverage of population. There is no clear relation between unit costs and service levels (quantity, quality, accessibility and reliability) between zones and technologies. The analysis suggests that allocations towards capital maintenance could help in reducing the gap between normative and observed lifespans.

While the approach of unconditional allocations towards provision of water in rural areas may be easier administratively and might benefit the low-cost regions, it would result in a less than desirable level outcome in the high-cost regions. There is need for rethinking on the policy of blanket or uniform allocations across the zones on the basis of the norms fixed at the state level. Added to this are the intra-village variations across socio-economic groups and geographical locations.

Multi-village schemes are not necessarily the best available option. In fact, single-village schemes appear to be more efficient despite all their drawbacks. One reason for this could be that the operation and management of multi-village schemes is split between contractors and the village panchayat. The village panchayat does not have the control over the quantum of water released and the time of release. On the other hand, the village panchayat is in full control of the system

Table 3.1 Cost per unit of water

Technology	Annualised cost /Cap (US$)		Annualised cost (US$)[a]		Service in quantity (lpcd)[b]	Population covered	Total amount of water used (m^3/year)	Cost per unit of water (US$/m^3)	
	Normative	Observed	Normative	Observed				Normative	Observed
HP	4.2	5.4	29,518	37,951	40	7,028	102,609	0.29	0.37
MPWS	6.9	16	59,575	138,144	42	8,634	132,359	0.45	1.04
SVS	6.6	15.6	229,178	541,694	41	34,724	519,645	0.44	1.04
MVS	7.0	17	140,966	342,346	41	20,138	301,365	0.47	1.14

Note [a]Costs are unit costs (fixed + recurring) as calculated earlier multiplied by the population covered under the technology
[b]These quantities are weighted average of summer and non-summer water use

in the case of single-village schemes. Though the management problems at the village level are same for both the schemes, SVS are plagued with the additional problems associated with source sustainability, water quality, etc.

It would be better to address these issues and strengthen the SVS rather than moving towards multi-village schemes, which are not efficient. What is more, MVS also will have source sustainability problems associated with climate change (IPCC 2008). In either case, source sustainability needs to be addressed effectively and management becomes easier in the context of single-village schemes with better planning. Further, there is need to revise the allocations to the sector in terms of magnitude and composition in the lines suggested here. LCCA is one tool that can help in achieving water security at the household level through judicious allocations towards source sustainability or source protection, water quality, capital maintenance, etc. It facilitates comprehensive planning with a pragmatic and integrated water resource management approach to rural water service delivery.

3.3 Good Practices of LCCA

As we have seen from the case studies, adoption in these studies or cases is not comprehensive enough to address various dimensions of sustainable development or green economy. This is mainly due to the difficulties in obtaining information on various aspects. This is more so in the case of environmental impacts. Only the recent studies have been more inclusive in this regard, though they too assess the impacts in the restricted boundaries. On the other hand, no studies deal with all three aspects of sustainability—economic, environmental and social. Addressing all three aspects is clearly a difficult task to achieve given the complexities of assessing all three aspects, and hence, finding an ideal LCCA application is a challenge practically. In this regard, good practices could be identified as those that follow the processes of LCCA.

The first step in adopting LCCA is to scope the possible linkages in the production process of a selected system. These should include economic, environmental and social aspects. Define the boundaries in which the assessment can be made scientifically and realistically given the data and other constraints. Though it may not be possible to assess the impacts monetarily in some cases, it would be good to mention those aspects and their expected impacts on the outcomes, positive or negative. This is critical from the social dimension, as these impacts need to be conveyed to the main stakeholders in the process.

Providing the structure of costs and their components is the next important step. Life-cycle costs are broadly divided into acquisition and sustaining costs. Identifying different components of these costs is a challenge. Barringer (1998) provided an 11-step process for identifying and including all the appropriate costs under the two broad cost categories. Here, we present one such detailed process followed in a case study from Singapore where a whole life-cycle cost approach was adopted (Sreenivasan 2013). It may be noted that here 'good practice' does not mean comprehensive or all-inclusive approach but limited to following a process.

The approach was followed in order to deliver life-cycle replacement on the New ITE College West, Singapore. This case study provides the process followed for the New ITE College. The objective of the assessment was to offer 'value for money' through optimum combination of whole life cost and quality to meet the user requirements. This approach helped to produce integrated-design solutions.

Following an integrated approach is critical in the process. This integration was attained between construction contractor, facilities manager and life-cycle advisor, which can ensure an integrated solution. This also helps in checking double counting and demonstrating value for money strategy. The key elements are the quantities and the prices provided by the cost planner. Besides, risk, uplift for work in the existing building, management fees and design fees for M&E work are taken into consideration. Replacement cycles are determined through supplier warranties, in house expertise, benchmarking and published data.

This approach is expected to (1) bring in best whole life value for money through numerous component options; (2) reduce risk by in-depth analysis of future costs; (3) ensure a competitive price via supply chain expertise and feedback from existing concessions; (4) provide optimum solutions and avoid any double counting; (5) ensure minimal disruption; (6) offer technological upgrades specifically planned and priced; and (7) achieve pricing transparency.

3.4 Constraints and Challenges in the Application of LCCA

Despite its importance, application of LCCA remains limited and is mostly undertaken at the early stages of procurement (Clift and Bourke 1999) and limited to construction (Wilkinson 1996; Sterner 2000). The reasons for this are practical as well as political. Often, capital costs and operating expenditure are met by different parties and there is no incentive on behalf of those responsible for construction to reduce the subsequent costs-in-use (Bull 1993).

Two other important reasons are the shortage of LCCA data and the complexity of the LCCA exercise. One of the main reasons for this is the lack of any frameworks or mechanisms for collecting and storing the data (Clift and Bourke 1999). The estimation of the life-cycle costs itself is too complex to calculate manually. Further, fixing appropriate discount rate is also a complex issue, as the discount rates could be different for different components depending on its nature—environmentally or socially beneficial processes or technologies need to apply lower rates of discount. All the costs must be discounted, added up and projected over the building's life cycle for each alternative design. Complex interrelations between different types of costs and elements might make it difficult to select the best possible option, as improvements in one area might have negative effects in others (Bakis et al. 2003). Some of these shortcomings are critical and serious in nature to the extent of questioning the validity and practicability of LCCA approach. At the same time, the advantage of the comprehensive approach is increasingly realised and the way forward is to work towards reducing the limitations of the approach.

Environmental issues are increasingly gaining attention of policymakers in the developing countries, though they are yet to get into the top priority list. Political economy factors constrain the promotion of environmental issues as apriority. As a result, environmental issues are often pushed through 'command and control regulation' policy instruments. The experience with the implementation of these command and control instruments has not been encouraging in the absence of complementary inter-sectoral policies. Of late, voluntary approaches are being considered as effective policy instruments to compliment the traditional command and control measures (Iraldo et al. 2014). The increasing demand from consumers for environmentally safe products and services is pushing the industry to address environmental issues voluntarily. Others include the use of incentive and disincentive structures for promoting or polluting environment and through negotiated agreements with private sector.

There is an urgent need to promote environmental issues in developing countries. Some of the environmental impacts are clearly resulting in unsustainable and irreversible damages (e.g. water, forestry and other common pool resources). Climate change impacts have further hastened the process of degradation. The degradation of resources coupled with the interlinkages between different sectors is resulting in strident constraints on basic amenities such as water, sanitation and power. And they are directly affecting the food security in developing countries, especially vulnerable regions such as rainfed areas. The linkages between unsustainable resource-use patterns and the sustainability of basic amenities and food security are only vaguely understood at the policy level. At the same time, unsustainable service delivery of basic amenities and unstable food security are putting pressure on policymakers to improve services and promote sustainable resource-use pattern. Hither to, the policy reactions to the problems have been in the nature of managing the problems in the short run rather than solving the problems in the end. This requires a systematic and scientific approach with judicious planning.

The development experience so far has been that issues or problems are taken up or solved in isolation. Given the interconnectedness of different sectors or subsectors within a particular sector, there is need for a systems approach. In most developing countries, there are no guidelines for project appraisal. In fact, in the case of public infrastructure projects, project appraisals are hardly carried out, though ex-post-evaluations are most common. Over the last decade or so environmental impact assessments are being made mandatory in large-scale projects (public as well as private) such as irrigation, mining and power. Of late, environmental or natural resource impacts find place in ex-post-evaluation of public-funded projects such as watershed development. But they are not comprehensive enough to incorporate environmental sustainability issues. One reason is that there are no guidelines on how to go about environmental impact assessments, though they are mandatory for getting approvals. As a result, environmental impact assessments are carried out as a formality rather than to achieve any objective(s) (say sustainable development).

The result is that the appraisals or evaluations remain partial in terms of addressing the interconnected issues and keep shifting the problem from one sector to another. As revealed in this review, LCCA is one of the most comprehensive tools used to assess the environmental impacts of a product or service. LCCA can be used to compare different technologies not only on their financial or economic merits, but also on their impacts on environment or natural resources. Combining economic and environmental impacts provide net returns to the technology. This provides the basis for selecting sustainable technologies/products/services. Moreover, it is shown that adoption of LCCA is also capable of ensuring sustainable services and food security. This could be achieved not only due to the interlinkages between basic services and natural environment but also due to its approach to costing.

The merits of LCCA in addressing environmental impacts are well recognised at the international level. Following the UN life-cycle thinking initiative, number of European countries has initiated policy commitments to adopt LCCA (Finnveden et al. 2009). Its adoption in developing countries is yet to take shape. Apart from low priority for environment at the policy level, awareness about LCCA itself is very limited. The adoption of LCCA in the private sector is also quite low in the absence of any policy guidance or regulations. At the same time, there is increasing awareness about environmental issues among consumers though the demand for such goods and services is quite limited due to high environmental premiums (organic foods).

How well do the so-called environmentally safe goods and services (at the consumer level) really contribute to green economy? It is observed that excess use of manure in the System of Rice Intensification (SRI) would increase methane emissions and greenhouse gases (Gathorne-Hardy 2013a, b). While SRI is being promoted for its water saving qualities (less water per kg of paddy produced), its other impacts are not well understood. For instance, the water saved in SRI is often used to expand the area under crops in the same location. When taken at the basin scale, there will not be any water savings for environmental requirements (environmental flows). Besides, SRI does not have any return flows (which is the case in flood irrigation) and hence reduces the availability of water downstream resulting in reduced environmental flows and inequity. This is observed even in the case of other water saving technologies (WCTs) (Batchelor et al. 2014).

Another case where such granularity is missed is wastewater usage for productive purposes. While wastewater is often let out into streams, ponds and rivers without treating it, its usage downstream for productive purposes not only creates jobs and income but also results in adverse health impacts. Unless the net impacts (positive-negative) are assessed, the economics of wastewater use would not be clear for making investment decisions to create infrastructure for wastewater treatment (Reddy and Kurian 2010). That is water sector policies and investment decisions should shift from single-use infrastructure to multiple-use infrastructure investment decisions. Such contradictions are also observed in the case of different biofuel production processes (Davis et al. 2008). Therefore, it is necessary to understand and adopt a comprehensive approach in order to move towards sustainable development. And macropolicy has a critical role in promoting such approaches and awareness in public as well as private organisations.

Given the fact that sustainable services and food security are integral to LCCA, adoption of LCCA could provide double benefits in developing countries, where dwindling services is a major policy concern. In this regard, LCCA could provide cost-effective measures as a sector-financing tool for sector efficiency. Adopting LCCA to finance, the sector would help to get the unit costs right and the right balance of different cost components for sustainable service delivery. In the case of environmental issues, European countries have introduced standardisation processes through International Organisation of Standards (ISO). ISO has developed standard labelling such as eco-labelling, environmental claims and eco-profiles for voluntary adoption (Iraldo et al. 2014). Even in Europe, the application of LCCA is limited to design stage and not applied in the implementation stage (Schiller and Dirlich 2013).

While adoption of LCCA provides win–win policy strategies in developing countries, there is need for awareness and capacity building for wider promotion and adoption of LCCA. LCCA is not a new concept in these countries although it needs recasting to address the present day concerns. LCCA is often viewed as an engineer's tool for project appraisal. Its evolution over the years as an effective tool to move towards sustainable development and service delivery has also encouraged planners and financial managers to adopt it with conviction across the world. This needs careful articulation in order to mainstream it into policymaking basically moving towards life-cycle thinking and life-cycle management of infrastructure projects. It is not to suggest that developing countries need to embark on the same path followed by the developed world. Understanding the potential and adoptability of LCCA to the local conditions, in terms of scale and intensity, is critical.

Apart from awareness and capacities, one of the main constraints in adopting LCCA in developing countries is the huge data requirements. LCCA is known for its data intensity and sensitivity to the methods and tools used in assessing the environmental impacts. Building on the data sources and ensuring data quality on various indicators across the sectors is a necessary first step. The most challenging aspect in this regard is the coordination between sectors and their departments for data generation and data sharing. For instance, inter-departmental coordination and integration has been on the cards for quite some time in countries like India, but yet to be implemented in practice. Creating information and feedback loops between the departments through centralised information system might help in overcoming this problem. Often important environmental data are not accessible to researchers or public though it is collected by the industry thus keeping the likely environmental impacts in the dark.

3.5 Conclusions

This volume on LCCA was aimed at enhancing the capacities of policymakers and practitioners with a view to identify potential aspects for its adaption in developing countries. The learning material is expected to influence the policy understanding of why LCCA is central to achieving the objectives of sustainable development as well as sustainable service delivery and to influence the behaviour of sector stakeholders. The broad objective is that LCCA is mainstreamed into governance processes at all institutional levels from local to national in order to increase the ability and willingness of the decision-makers (both users and those involved in service planning, budgeting and delivery) to make informed and relevant choices between different types and levels of products and services.

This volume, based on the experience of earlier studies and selected case studies, argues that a comprehensive LCCA can provide 'win–win' strategies in terms of identifying appropriate technologies, products and services that are environmentally, economically and socially sustainable. LCCA prompts policy shifts towards broader and systems perspective. LCCA is not only a tool that can be used in policy planning as and when necessary. Adoption of LCCA evolves from life-cycle thinking that needs to be ingrained into the macropolicy. LCCA management processes need to be put in place. This calls for awareness building and capacities at the policy and planning levels. Here, we provide the key merits of LCCA that can attract quick policy attention in developing countries.

1. LCCA is an appraisal tool that can be applied at any stage of the life cycle. This helps in evaluating even the existing infrastructure investments.
2. LCCA has the potential to deal with the nexus approach by adopting a systems approach that includes inter-sectoral linkages and externalities.
3. LCCA is now widely used covering most of the sectors, products and services. Common or standard LCCA guidelines can help in following a systematic economy-wide approach.
4. LCCA can ensure sustainable services through its comprehensive approach dealing with all aspects of nexus.
5. LCCA can be used as a budgeting tool, which can ensure allocations towards source sustainability, asset management, natural disasters, etc. This provides the much-needed sustainability of services.

Adoption of LCCA as a budgeting tool needs to be taken up at the national- and state-level budgeting processes. There is need for more research in the context of developing countries to establish and convince policymakers in this regard. Action research on the adoption of LCCA in some key sectors would be a good starting point in this direction.

Keywords and Definitions

Data mining	'The process of analysing data from different perspectives and summarising it into useful information. Technically, data mining is the process of finding correlations or patterns among dozens of fields in large relational databases'
Data quality	Characteristics of data that relate to their ability to satisfy stated requirements
Data set (LCI or LCIA data set)	Document or file with life-cycle information of a specified product or other reference (e.g. site, process), covering descriptive metadata and quantitative life-cycle inventory and/or life-cycle impact assessment data, respectively (European Commission—Joint Research Centre-Institute for Environment and Sustainability)
Environmental life-cycle assessment (E-LCA)	E-LCA 'is an assessment technique that aims at addressing the environmental aspects and their potential environmental impacts throughout a product's life cycle'
Environmental aspect	Element of an organisation's activities, products or services that can interact with the environment
Environmental product declaration (EPD)	'An EPD is a standardised (ISO 14025/TR) and LCA-based tool to communicate the environmental performance of a product or system and is applicable worldwide for all interested companies and organisations' (http://www.environmentalproductdeclarations.com/)
Impact category	'Impact categories are logical groupings of life-cycle assessment results of interest to stakeholders and decision-makers'
Life cycle	'Consecutive and interlinked stages of a product system, from raw material acquisition or generation from natural resources to final disposal'
Life-cycle approaches	'Techniques and tools to inventory and assess the impacts along the life cycle of products'
Life-cycle assessment (LCA)	'Compilation and evaluation of the inputs, outputs and the potential environmental impacts of a product system throughout its life cycle'

Life-cycle costing (LCC)	'Life-cycle costing, or LCC, is a compilation and assessment of all costs related to a product, over its entire life cycle, from production to use, maintenance and disposal'
Life-cycle data set library	'A global database of registered and searchable life-cycle data sets'
Life-cycle impact assessment (LCIA)	'The phase of life-cycle assessment aimed at understanding and evaluating the magnitude and significance of the potential environmental impacts for a product system throughout the life cycle of the product'
Life-cycle interpretation	'The phase of life-cycle assessment in which the findings of either the inventory analysis or the impact assessment, or both, are evaluated in relation to the defined goal and scope in order to reach conclusions and recommendations'
Life-cycle inventory (LCI)	'The phase of life-cycle assessment where data are collected, the systems are modelled, and the LCI results are obtained'
Life-cycle inventory analysis	'The phase of life-cycle assessment involving the compilation and quantification of inputs and outputs for a product throughout its life cycle'
Life-cycle inventory database	'A system intended to organise, store, and retrieve large amounts of digital LCI data sets easily. It consists of an organised collection of LCI data sets that completely or partially conforms to a common set of criteria, including methodology, format, review and nomenclature and that allows for interconnection of individual data sets that can be specified for use with identified impact assessment methods in application of life-cycle assessments and life-cycle impact assessments'
Life-cycle management (LCM)	'A product management system aimed at minimising the environmental and socio-economic burdens associated with an organisation's product or product portfolio during its entire life cycle and value chain. LCM supports the business assimilation of product policies adopted by governments. This is done by making life-cycle approaches operational and through the continuous improvement of product systems'

Life-cycle management systems	'Management systems that incorporate the basic life-cycle principles plus key elements of ISO 9000, ISO 14000 and ISO 26000 to ensure continuous improvement: the plan-do-check-act cycle; policy, objectives and targets; procedures and instructions; monitoring and registration systems; and documentation and reporting'
Life-cycle programmes and activities	'Initiatives that support decision-making based on life-cycle thinking at one or more organisational units (e.g. at the design, procurement, recycling units)'
Life-cycle public policies	Public policies that incorporate or are based on life-cycle guiding principles
Life-cycle sustainability assessment (LCSA)	Evaluation of all environmental, social and economic negative impacts and benefits in decision-making processes towards more sustainable products throughout their life cycle
Social life-cycle assessment (S-LCA)	'A social and socio-economic life-cycle assessment (S-LCA) is a social impact (real and potential impacts) assessment technique that aims to assess the social and socio-economic aspects of products and their positive and negative impacts along their life-cycle encompassing extraction and processing of raw materials; manufacturing; distribution; use; reuse; maintenance; recycling; and final disposal'
Sustainable consumption and production	'The use of services and related products, which respond to basic needs and bring a better quality of life while minimising the use of natural resources and toxic materials as well as the emissions of waste and pollutants over the life cycle of the service or products so as not to jeopardise the needs of future generations'

References

Bakis N, Kagiouglou M, Aouad G, Amaratunga D (2003) An integrated environment for life-cycle costing in construction. Constr Inform Digital Libr. http://itc.scix.net/paperw78-2003-15.content. Retrieved 10 Apr 2014

Barringer HP (1998) Life cycle cost and good practices. In: Presented at the NPRA maintenance conference, San Antonio Convention Center, San Antonio, Texas, May 19–22

Batchelor C et al (2014) Do water-saving technologies improve environmental flows? J Hydrol

Bull JW (1993) The way ahead for life cycle costing in the construction industry. In: Bull JW
    (ed) Life cycle costing for construction. Blackie Academic & Professional, Glasgow
Clift M, Bourke K (1999) Study on whole life costing. BRE report 367. CRC, Boca Raton
Davis SC, Anderson-Teixeira, KJ, DeLucia EH (2008) Life-cycle analysis and
    the ecology of biofuels. Trends Plant Sci 14(3). Retrieved 10 Apr 2014 from
    http://www.life.illinois.edu/delucia/Publications/Davis%20Life%20Cycle.pdf.
    doi:10.1016/j.tplants.2008.12.006
Finnveden G et al (2009) Recent developments in life cycle assessment. J Environ Manag doi:
    10.1016/j.jenvman.2009.06.018
Gathorne-Hardy A (2013a) Baselines and boundaries for rice LCA. In: International symposium
    on technology, jobs and a lower carbon future: methods, substance and ideas for the informal
    economy (the case of rice in India), 13–14 June 2013. Organised by University of Oxford
    and Institute of Human Development, India International Centre, New Delhi. Retrieved 10
    Apr 2014 from http://www.southasia.ox.ac.uk/sites/sias/files/documents/Conference%20
    Book.pdf
Gathorne-Hardy A (2013b) A life cycle assessment of four rice production systems: high yield-
    ing varieties, rain-fed rice, system of rice intensification and organic rice. In: International
    symposium on technology, jobs and a lower carbon future: methods, substance and ideas for
    the informal economy (the case of rice in India), 13–14 June 2013. Organised by University
    of Oxford and Institute of Human Development, India International Centre, New Delhi.
    Retrieved 10 April 2014 from http://www.southasia.ox.ac.uk/sites/sias/files/documents/
    Conference%20Book.pdf
IPCC (2008) Climate change and water, Technical paper VI, Inter Governamental Panel on
    Climate Change, Geneva
Iraldo F, Testa F, Bartolozzi I (2014) An application of life cycle assessment (LCA) as a green
    marketing tool for agricultural products: the case of extra-virgin olive oil in Val di Cornia,
    Italy. J Environ Planning Manage 57(1):78–103. doi:10.1080/09640568.2012.735991
Reddy VR, Jayakumar N, Venkataswamy M, Snehalatha M, Batchelor C (2012) Life-cycle costs
    approach (LCCA) for sustainable water service delivery: a study in rural Andhra Pradesh,
    India. J Water, Sanitation Hygiene Dev 02:279–290
Reddy VR, Kurian M (2010) Approaches to economic and environmental valuation of domestic
    wastewater. In: Kurian M, McCarney P (eds) Peri-urban water and sanitation services: Policy,
    planning and method. Springer, Dordrecht, pp 213–242
Schiller G, Dirlich S (2013) Applications of LCC for design and implementation of water and
    wastewater projects in Europe. Discussion paper presented at the NEXUS observatory work-
    shop on life-cycle cost assessment of infrastructure projects, UNU-FLORES and TERIU,
    New Delhi, 18–19 Dec 2013
Sreenivasan S (2013) Lifecycle case study ITE New College West, Singapore. Discussion paper
    presented at the NEXUS observatory workshop on life-cycle cost assessment of infrastruc-
    ture projects, UNU-FLORES and TERIU, New Delhi, 18–19 Dec 2013
Sterner E (2000) Life-cycle costing and its use in the Swedish building sector. Building Res Inf
    28(5/6):387–393
Wilkinson S (1996) Barriers to LCC use in the New Zealand construction industry. In: 7th inter-
    national symposium on economic management of innovation, productivity and quality in
    construction, Zagreb, pp 447–456